Marion Baumann

Discrete Time Analysis of Multi-Queue Systems with Multiple Departure Streams in Material Handling and Production under Different Service Rules

WISSENSCHAFTLICHE BERICHTE

Institut für Fördertechnik und Logistiksysteme
am Karlsruher Institut für Technologie (KIT)

BAND 94

Discrete Time Analysis of Multi-Queue Systems with Multiple Departure Streams in Material Handling and Production under Different Service Rules

by

Marion Baumann

KIT Scientific Publishing

Karlsruher Institut für Technologie
Institut für Fördertechnik und Logistiksysteme

Discrete Time Analysis of Multi-Queue Systems
with Multiple Departure Streams in Material Handling
and Production under Different Service Rules

Zur Erlangung des akademischen Grades einer Doktor-Ingenieurin
von der KIT-Fakultät für Maschinenbau des Karlsruher Instituts für
Technologie (KIT) genehmigte Dissertation

von M.Sc. Marion Baumann

Tag der mündlichen Prüfung: 7. Juni 2019
Referent: Prof. Dr.-Ing. Kai Furmans
Korreferent: Prof. Dr. George Liberopoulos

Impressum

KIT Scientific Publishing

Karlsruher Institut für Technologie (KIT)
KIT Scientific Publishing
Straße am Forum 2
D-76131 Karlsruhe

KIT Scientific Publishing is a registered trademark
of Karlsruhe Institute of Technology.
Reprint using the book cover is not allowed.

www.ksp.kit.edu

Print on Demand 2019 – Gedruckt auf FSC-zertifiziertem Papier

ISSN 0171-2772
ISBN 978-3-7315-0984-4
DOI 10.5445/KSP/1000098922

Discrete Time Analysis of Multi-Queue Systems with Multiple Departure Streams in Material Handling and Production under Different Service Rules

Zur Erlangung des akademischen Grades

Doktor der Ingenieurwissenschaften (Dr.-Ing.)

der KIT-Fakultät für Maschinenbau
Karlsruher Institut für Technologie (KIT)
genehmigte

Dissertation

von

M.Sc. Marion Baumann

Tag der mündlichen Prüfung:	07. Juni 2019
Referent:	Prof. Dr.-Ing. Kai Furmans
Korreferent:	Prof. Dr. George Liberopoulos

Danksagung

Die vorliegende Arbeit entstand während meiner Tätigkeit als wissenschaftliche Mitarbeiterin am Institut für Fördertechnik und Logistiksysteme des Karlsruher Instituts für Technologie (KIT). Ich möchte mich an dieser Stelle bei allen Personen bedanken, die zum Gelingen dieser Arbeit beigetragen haben.

Ich bedanke mich bei meinem Doktorvater Prof. Dr.-Ing. Kai Furmans, Leiter des Instituts für Fördertechnik und Logistiksysteme, für die Übernahme des Hauptreferats. Seine Forschungsarbeiten im Bereich der zeitdiskreten bedientheoretischen Methoden haben mich zu dieser Arbeit motiviert. Die Möglichkeit zum freien und selbstständigen Arbeiten, zusammen mit kreativen und zielführenden Diskussionen, haben zum erfolgreichen Abschluss meiner Dissertation beigetragen.

Prof. Dr. George Liberopoulos danke ich für die Übernahme des Koreferats sowie die Möglichkeit acht Wochen bei ihm am Department of Mechanical Engineering an der University of Thessaly in Griechenland zu forschen. Während dieser Zeit konnte ich mich ausschließlich meiner wissenschaftlichen Arbeit widmen.

Bei Prof. Dr.-Ing. Marcus Geimer bedanke ich mich für die Übernahme des Prüfungsvorsitzes meiner Promotionsprüfung.

Allen aktiven und ehemaligen Kollegen danke ich für die angenehme Arbeitsatmosphäre am Institut und die gegenseitige Unterstützung.

Insbesondere möchte ich mich bei Dr.-Ing. Melanie Schwab bedanken, die meine Begeisterung für das wissenschaftliche Arbeiten während meiner Zeit als studentische Hilfskraft und bei meiner Masterarbeit geweckt hat. Dr.-Ing. Martin Epp, der mich in den ersten Jahren am Institut begleitet und geführt hat, danke ich für die thematische Unterstützung und die anregenden Diskussionen. Außerdem danke ich Dr.-Ing. Katharina Fleischer-Dörr, die mich immer wieder motiviert und unterstützt hat. Dr.-Ing Maximilian Hochstein danke ich für den gemeinsamen Endspurt, währenddessen wir uns gegenseitig mit Mut und Zuspruch gestärkt haben.

Ein großer Dank gilt meiner Familie und all meinen Freunden, die immer an mich geglaubt haben. Insbesondere danke ich meinen Eltern, die mich auf meinem eingeschlagenen Lebens- und Bildungsweg stets unterstützt und gefördert haben. Ganz besonderer Dank gilt meiner Mutter Helena Rimmele, die mir immer mit einem offenen Ohr und einem guten Rat zur Seite stand und beim Korrekturlesen viele Tage und Nächte geopfert hat.

Mein allergrößter Dank gilt meinem Mann Matthias Baumann für seine grenzenlose Unterstützung. Er hat mich während der gesamten Promotion ermutigt weiterzumachen und sich selbst dabei häufig zurückgenommen.

Karlsruhe, September 2019 Marion Baumann

Kurzfassung

Bedienstrategien werden in verschiedenen Forschungs- und Anwendungsbereichen verwendet. Sie werden eingesetzt, wenn mehrere Kunden wie zum Beispiel Aufträge, Fördereinheiten oder Nachrichten von einer Ressource bedient werden sollen. Durch die unterschiedlichen Forschungs- und Anwendungsbereiche gibt es eine Vielzahl von Klassifizierungen, Modellen und Auswertungen der Bedienstrategien bezogen auf den jeweiligen Bereich. Die Klassifikationen, Untersuchungen und Auswertungen von Bedienstrategien beziehen sich jedoch in der Regel nur auf einen bestimmten Bereich. Die in der Literatur entwickelten Modelle zur Abbildung von Bedienstrategien basieren auf vereinfachten Annahmen, die in der Regel nicht zutreffen. Es fehlt ein ganzheitliches Modell für verschiedene Bedienstrategien, das die Verteilungen der Leistungsparameter ohne restriktive Annahmen bestimmt.

Ziel dieser Arbeit ist es, einen zeitdiskreten Modellierungsansatz zu entwickeln, um verschiedene Bedienstrategien ganzheitlich abzubilden. Das entwickelte Modell heißt *Multi-Bediensystem mit mehreren Ausgangsströmen (MQSMDS)*. Aus der Analyse und Bewertung basierend auf dem Modell können Empfehlungen für die sinnvolle Anwendung der Bedienstrategien in einem breiten Spektrum von Forschungs- und Anwendungsbereichen abgeleitet werden. Mit den Ergebnissen dieser Arbeit ist eine schnelle und kostengünstige Analyse und Modellierung bestehender und geplanter spezifischer Förder- und Produktionssysteme sowie eine schnelle und einfache Identifizierung geeigneter Bedienstrategien für diese Systeme möglich.

In dieser Arbeit wird eine ganzheitliche Klassifikation bestehend aus zwei Regelkategorien, sieben Regelklassen und 16 Regelarten basierend auf der Literatur aus den verschiedenen Forschungs- und Anwendungsbereichen erstellt. Aus der Kombination der verschiedenen Regelarten ergeben sich insgesamt 480 Bedienstrategien, die mit der Klassifizierung modelliert werden können. Das Modell des MQSMDS ist als zeitdiskrete Markov-Kette modelliert und die Verteilungen der Leistungsparameter wie zum Beispiel die Verteilung der Anzahl der wartenden Kunden und der gesamten Durchlaufzeit werden berechnet. Auf der Grundlage der numerischen Bewertung der Bedienstrategien werden Empfehlungen für die geeignete Anwendung der Bedienstrategien ausgesprochen. Durch die Untersuchung der Systemeigenschaften in Abhängigkeit von den Systemparametern und der gewählten Bedienstrategie kann die Anzahl der relevanten Bedienstrategien von 480 auf zehn reduziert werden. In Abhängigkeit von den Einflussgrößen und dem angestrebten Optimierungsziel werden die entsprechenden Bedienstrategien festgelegt. So müssen bei der Auswahl einer Bedienstrategie für ein System nur wenige Bedienstrategien im Detail betrachtet und im Einzelfall gegenübergestellt werden.

Abstract

Service rules are applied in various research and application areas. They are used when several customers like jobs, conveying units or messages want to be served by one resource. Due to the various research and application areas, there is a large number of classifications, models and evaluations of the service rules related to the specific area. However, the classifications, the investigations and the evaluations of service rules usually only refer to a specific area. The models developed from the literature that depict service rules are based on simplified assumptions that generally do not apply. A holistic model for different service rules that determines the performance parameter distributions without restrictive assumptions is missing.

The objective of this thesis is to develop a modelling approach in discrete time domain in order to depict different service rules holistically. The developed model is called *multi-queue system with multiple departure streams (MQSMDS)*. The analysis and evaluations based on the model can be used to make recommendations about the appropriate use of the service rules in a wide range of research and application areas. With the results of this work a rapid and low-cost analysis and modelling of existing and planned specific material handling and production systems as well as a fast and easy identification of suitable service rules for these systems is possible.

In this work a holistic classification consisting of two rule categories, seven rule classes and 16 rule types is created based on the literature from the various research and application areas. The combination of the different rule types results in a total of 480 service rules that can be modelled with the classification. The MQSMDS is modelled as a discrete time Markov chain and the distributions of the performance parameters such as the distributions of the number of waiting customers and the total sojourn time are calculated. On the basis of the numerical evaluation of the service rules, recommendations are made about the appropriate use of the service rules. By examining the system characteristics depending on the system parameters and the selected service rule, the number of relevant service rules can be reduced from 480 to ten. The appropriate service rules depending on the influence parameters and the optimization objective pursued are determined. Thus, when choosing a service rule for a system, only a few different service rules have to be examined in detail and compared for the individual case.

Contents

1 Introduction

Any fool can make a rule,
and any fool will mind it.
-Henry David Thoreau

In complex situations simple rules help to make a good decision. They support us to structure and ease our everyday life. Especially for children fixed rules are very important. When the traffic light shows red, we stop. Before we go to bed, we clean our teeth. These basic rules you learn as a child. As an adult, we follow rules, which secure our social coexistence and a good life or which are meant to help us to live healthy. For this last purpose various fitness tracers establish the rule that you should make at least 10,000 steps daily if you want to promote your health in the long term, improve your daily performance and prevent various chronic diseases. But this number is a completely arbitrary figure. There are no scientific studies which justify this special number. It is solely based on a marketing campaign from 1964, when a Japanese manufacturer released a step-counter during the Olympic Games in Tokyo which was called 'The 10,000 Step Counter'. This round and easily remembered number convinced health organizations and they propagate this value ever since (Cox 2018). This example shows that universally accepted rules may not necessarily have reliable scientific evidence and it may be required to critically question their basis. But nonetheless the rule of the 10,000 steps enhances the motivation to move even though the number of the steps is selected randomly.

The sense of rules in road traffic is obvious. In addition to the basic rule of mutual consideration (§1 StVO), one of the most important regulations is that of priority (§8 StVO). In Germany, at junctions, whoever comes from the right has right of way, except the right of way is regulated by traffic signs. The traffic rules serve the purpose of safety, the flow of traffic and they help to clarify the question of fault in accidents (Janker 2018).

With this example, the properties of a rule can be easily derived:

1. A rule pursues an objective.
2. A rule is unique with no contradiction and an unambiguous result.
3. A rule is easy to understand.

The pursued objective plays a decisive role. Thus, a rule can be evaluated according to the achievement of the goal. However, the evaluation of a rule is not always easy, since the effects of the rule depend on certain circumstances. For example, the right before left rule at certain intersections can significantly improve safety. On the other hand, the rule can increase the accident rate on a frequented road with an unclear intersection.

In material flow and production systems there are different rules that determine which order will be processed next. Because of their simplicity, they are often preferred to complex optimization methods in operational practice. *First come, first served (FCFS)* synonymous with *first-in-first-out (FIFO)* is one of the most simple dispatching policies. Objectives that are pursued by the choice of a rule are utilization, performance, time saving, reduction of queueing or safety. These objectives are usually not compatible with each other and can not be achieved by the same rule, so the appropriate rule must be found depending on the application (Gudehus and Kotzab 2012).

Figure 1.1 shows an example with four orders that are available for processing. The orders have arrived in the sequence (A, B, C, D). If the orders are processed according to the FCFS rule, the average waiting time is ten minutes. If the *shortest job first (SJF)* rule is used, the average waiting time is reduced to four minutes. On the other hand, with the *longest job first (LJF)* rule, the average waiting time increases to eleven minutes. However, the sojourn time of order *A* increases significantly with rule SJF, with the consequence that the due date may not be observed.

Figure 1.1: Example of the impact of service rules on waiting time

The purpose of this work is to introduce an analytical model which allows a quantitative analysis of service rules in queueing systems. The model can be used to investigate and compare different service rules. The presented insights into the system's behaviour can be used to improve material flow and production systems.

1.1 Problem Description

The research of this work can be divided into three parts with three research questions. In the following, the problem description for each part is presented and the respective research question is derived.

Service rules are applied in various research and application areas. They are used when several customers, like jobs, conveying units or messages, want to be served by one resource. Accordingly, many different rules have been developed for different research and application areas. But usually the service rules only refer to a specific area. For example, the service disciplines *exhaustive service*, *gated service* and *limited service* are investigated in so-called polling systems used to model multiplexer or a router with several inputs (Takagi and Kleinrock 1984). In production systems there are dispatching policies like *shortest job first*, *earliest due date* or *minimum slack time* (Blackstone et al. 1982). There is no holistic classification of service rules across the various research and application areas. In order to be able to carry out investigations related to service rules, however, the service rules to be considered must first be defined and classified. The following research question can be derived:

1. Which service rules can be generally applied?

There are different quantitative methods of performance analysis for the evaluation of material handling and production systems. Schleyer (2007) classifies the methods *continuous time-queueing analysis*, *discrete time queueing analysis* and *simulation* and names advantages of modelling in discrete time. The advantages of discrete time queueing analysis can be summarized as follows:

- Accuracy: The exact calculation of the distributions of the performance parameters and thus not only the expected values but also quantiles, enables a highly accurate analysis with respect to stochastic processes.
- Level of detail: Modelling with discrete input distributions allows the use of exact empirically determined data and thus a mapping of real processes with a high degree of accuracy.
- Efficiency: Discrete time queueing analysis has a low modelling effort and a short computing time for the determination of performance parameters compared to simulation.

A challenge in discrete time queueing models is the mapping of different streams. There is no model to represent flows from different customer classes that are served in the same system. For example, the sojourn time distribution of different product families in a production system cannot be calculated with the existing discrete time queueing models. Streams can be split by a stochastic split and merged again by a stochastic merge (Furmans 2004). Furmans et al. (2015) model a roller lift table by combining a stochastic merge, a G|G|1 queueing station and a stochastic split. Due to the assumed independence, this discrete time model is an approximation. Furthermore, the model is based on a *first come, first served (FCFS)* service rule. Other service rules cannot be depicted with the model.

Service rules are analysed using priority queueing models or polling models. Takagi (1996) presents a bibliography on polling models that contains over 700 publications. However, the models are often based on simplified assumptions such as a Poisson arrival process, symmetric systems and negligibly small switching times. Furthermore, the models often only determine expected values of performance parameters. Models in discrete time in which the distributions of the performance parameters are calculated are usually approximative. Additionally, in the literature, only one or a few service rules from the corresponding research or application area are considered. A holistic model for different service rules that determines the performance parameter distributions exactly without restrictive assumptions is missing. On this basis, a further research question can be derived:

2. **How can a discrete time queueing model with different service rules be modelled?**

The investigation and evaluation of service rules is usually based on a specific research or application area and includes one or a few service rules. Conway et al. (1967) examine the dispatching policy *shortest job first* and show that this policy leads to lower inventories, shorter mean lead times, lower mean delay per order and high delivery reliability. Boon and Adan (2009) develop a new service discipline *mixed gated/exhaustive service* and evaluate it in comparison to gated and exhaustive service. An evaluation and comparison of the service rules depending on the conditions of the system across the various research and application areas does not exist yet. Therefore the following research question can be derived:

3. Which service rules should be used under which conditions?

These three research questions lead to the objective of this thesis to develop a modelling approach in the discrete time domain for different service rules. The different service rules are classified independently of the various research and application areas. The developed model is called *multi-queue system with multiple departure streams (MQSMDS)*. Based on this model, the performance parameter distributions are calculated exactly in order to be able to determine and compare both expected values and the quantiles of the characteristic values. The analysis and evaluations based on the model can be used to make recommendations about the appropriate use of the service rules. The research project should thus enable the rapid and low-cost analysis and modelling of existing and planned specific material handling and production systems and to identify suitable service rules for the systems.

1.2 Organization of the Thesis

This thesis is divided into eight chapters. In Figure 1.2 the research questions from Section 1.1 are assigned to the chapters and the structure of the book is illustrated graphically.

First the motivation and the fundamentals of the work are presented. The introduction in Chapter 1 is followed by a description of the basics of discrete time queueing analysis, on which the developed model is based (see

Chapter 2). A literature review on different service rules in various research and application areas and the corresponding queueing models that are used for modelling the service rules is presented in Chapter 3.

In order to answer research question 1, the service rules are classified in Chapter 4 on the basis of the literature review. Two rule categories are defined which consist of different rule types. The resulting service rules are derived by combining the rule types from rule category 1 and 2 and assigned to the various service rules from the research and application areas.

In Chapter 5 the *multi-queue system with multiple departure streams (MQSMDS)* under different service rules is developed and research question 2 is answered. After a system description of the MQSMDS with the basic notations of the random variables, the MQSMDS is modelled as a discrete time Markov chain. The calculations of the performance parameters are derived from the Markov chain. Since the calculations with the analytical model are limited due to the large state space, a simulation model is developed and validated in Chapter 6.

The third research question is answered in Chapter 7. On the basis of a numerical evaluation the system characteristics are examined, the service rules are evaluated and recommendations of service rules depending on parameters and objectives are made.

In Chapter 8, the results are summarized in a conclusion and an outlook on further research questions is given.

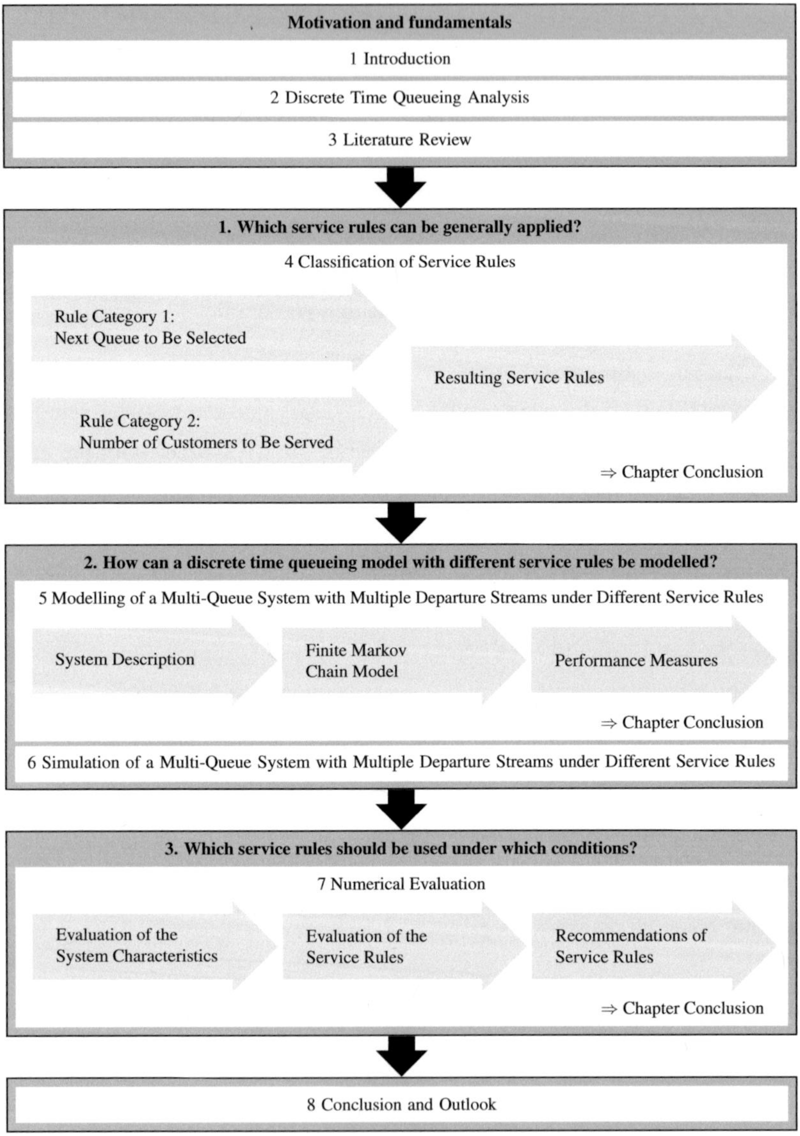

Figure 1.2: Organization of the book

2 Discrete Time Queueing Analysis

I think a simple rule of business is,
if you do the things that are easier first,
then you can actually make a lot of progress.
-Mark Zuckerberg

In this chapter, fundamental aspects of discrete time probability theory are presented. The explanations on the statistical fundamentals are based on Feller (1970) and Bol (2007). The theoretical foundations of stochastic models and queueing theory have been derived from Bolch et al. (2006), Kleinrock (1975), Tran-Gia (2005) and Waldmann and Stocker (2013).

In Section 2.1 the basics of discrete time probability theory are discussed, on which the mathematical model developed in this work is based. Subsequently, in Section 2.2 the Markov chain is described as a discrete time stochastic process and its properties. Finally, Section 2.3 introduces methods and models in the discrete time domain.

2.1 Basics of Discrete Time Probability Theory

The main characteristic of discrete time modelling is a discretization of the time into constant intervals of the length t_{inc}. It is assumed that all processes whose duration is an integer multiple of t_{inc} start synchronously. As a result, the observation of the system can be limited to integer multiples of t_{inc}.

Events are described by a discrete random variable X, whose distribution called probability mass, is defined as follows:

$$P(X = i \cdot t_{inc}) = x_i \qquad \forall\, i = 0, \ldots, i_{max} \tag{2.1}$$

Since t_{inc} is a constant the distribution can be shortened to:

$$P(X = i) = x_i \qquad \forall\, i = 0, \ldots, i_{max} \tag{2.2}$$

Since it can be assumed that all processes are finite, the definition range of X can be restricted to the interval $[0, i_{max}]$. The vector $\vec{x}$ is defined as the probability vector of the distribution X:

$$\vec{x} = \begin{pmatrix} x_0 \\ \vdots \\ x_{i_{max}} \end{pmatrix} \tag{2.3}$$

The distribution function of X results from summing all probabilities which are less than or equal to a value of i:

$$P(X \leq i) = \sum_{j=0}^{i} x_j \qquad \forall\, i = 0, \ldots, i_{max} \tag{2.4}$$

The expected value of X can be defined as follows:

$$E(X) = \sum_{i=0}^{i_{max}} i \cdot x_i \tag{2.5}$$

The variance can be used to make a statement about the dispersion of the probability distribution within the definition range of X:

$$Var(X) = \sum_{i=0}^{i_{max}} (i - E(X))^2 \cdot x_i = E(X^2) - (E(X))^2 \tag{2.6}$$

The squared coefficient of variation relates the variance to the expected value and is thus interpreted as a normalized measure of dispersion:

$$c_X^2 = \frac{Var(X)}{E(X)^2} \tag{2.7}$$

For the performance evaluation of a material handling or production system, it is interesting to know, whether the system can achieve a required target parameter σ_u with a given probability u. If the distribution function of the random variable is known, this can be determined using the $u\%$-quantile σ_u:

$$\sigma_u \Leftrightarrow \left(\sum_{j=0}^{\sigma_u} x_j \geq u\right) \wedge \left(\sum_{j=0}^{\sigma_u - 1} x_j < u\right) \tag{2.8}$$

The sum of two independent non-negative random variables X and Y is the convolution of the two distributions. It can be calculated as follows:

$$z_i = \sum_{j=0}^{i} x_j \cdot y_{i-j} \qquad \forall i \in \{0, \ldots, x_{max} + y_{max}\} \tag{2.9}$$

Based on the convolution operator $\otimes$, the calculation of the convoluted distribution $\vec{z}$ can be represented as follows:

$$\vec{z} = \vec{x} \otimes \vec{y} \tag{2.10}$$

The convolution of a distribution with itself is indicated by the superscript operator similar to the power notation. The convolution of the distribution of a random variable can be expressed as follows:

$$\begin{aligned} \vec{x}^{\otimes 0} &= \vec{x} \\ \vec{x}^{\otimes 1} &= \vec{x} \otimes \vec{x} \\ \vec{x}^{\otimes 2} &= \vec{x} \otimes \vec{x} \otimes \vec{x} \end{aligned} \tag{2.11}$$

2.2 Modelling of a Discrete Time Markov Chain

A discrete time Markov chain (DTMC) is a stochastic process $(X_n)_{n\in\mathbb{N}_0}$ with countable state space I which has the Markov property. This means that the following applies to all times $n \in \mathbb{N}_0$ and all states $i_0,\ldots,i_n,i_{n+1} \in I$:

$$P(X_{n+1} = i_{n+1} \,|\, X_0 = i_0,\ldots,X_n = i_n) = P(X_{n+1} = i_{n+1} \,|\, X_n = i_n) \qquad (2.12)$$

The Markov property is also known as the memorylessness of the stochastic process, since the future development of the process depends exclusively on the last observed state and is independent of all previous states. The conditional probability $P(X_{n+1} = i_{n+1} \,|\, X_n = i_n)$, with which the Markov chain changes from state i_n to state i_{n+1}, is called the transition probability p_{ij}. If the transition probabilities are independent of the time n, then the Markov chain is homogeneous. The transition probabilities are combined to a transition matrix $P = (p_{ij})$ where the following applies:

$$\sum_{j\in I} p_{ij} = 1, \qquad p_{ij} \geq 0 \qquad \forall i,j \in I \qquad (2.13)$$

The state probabilities $\vec{\pi}(n)$ of the Markov chain at the time $n \in \mathbb{N}_0$ are calculated with an initial distribution $\vec{\pi}(0)$ and the transition probabilities:

$$\vec{\pi}(n) = \vec{\pi}(0) \cdot P^n = \vec{\pi}(n-1) \cdot P \qquad (2.14)$$

A distribution is called stationary, if the following applies to all $n \in \mathbb{N}_0$:

$$\pi_j = \sum_{i\in I} \pi_i \cdot p_{ij}^{(n)} \qquad \forall j \in I \qquad (2.15)$$

This means that in the steady state, the system state i is identically distributed at each observation time.

The steady state distribution $\vec{\pi}$ of the Markov chain is determined by solving the following system of linear equations:

$$\pi_j = \sum_{i \in I} \pi_i \cdot p_{ij} \tag{2.16}$$

In addition, the following normalization condition and non-negativity condition applies:

$$\sum_{i \in I} \pi_i = 1, \qquad \pi_i \geq 0 \qquad \forall i \in I \tag{2.17}$$

Linear equation systems can be solved by the Gauss algorithm. The linear system for determining the steady state distribution which is overdetermined due to the normalization condition can be solved by the Householder transformation. This procedure is described in Engeln-Müllges et al. (2011).

2.3 Queueing Models in Discrete Time

The classic queueing model is known as a single server queueing system. It consists of a queueing buffer of infinite size and one server. A server can only serve one customer at the same time. If all servers are busy when a customer arrives, the newly arriving customer is buffered and waits for service. When the service of the customer currently served is finished, one of the waiting customers will be selected for service based on a queueing discipline. Figure 2.1 schematically shows a single station queueing system with one server and an unlimited buffer size.

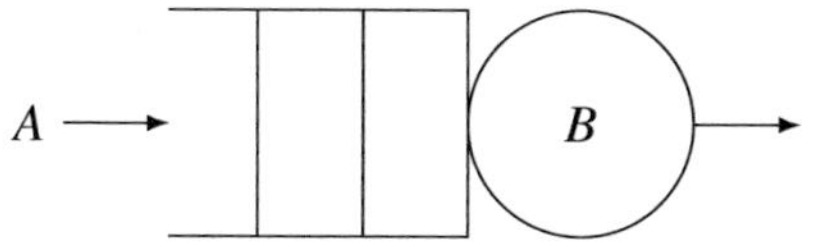

Figure 2.1: Singe station queueing systems

According to Kendall (1953), a queueing system can be described by the following properties:

- the arrival process,
- the service process,
- the waiting room capacity,
- the number of servers and
- the queueing discipline.

The queueing system is in equilibrium when the utilization is less than 1. The utilization ρ is calculated by dividing the arrival rate λ by the service rate μ. Alternatively, it can also be defined as the expected value of the proportion of a busy server. For a queueing system in a steady state, Little's Law (Little 1961) applies, which states that the number of customers in the system L corresponds to the arrival rate λ multiplied by the sojourn time W of a customer in the system.

In discrete time, the system is considered at equally spaced time epochs, with each discretization interval having a length of t_{inc} (see 2.1). As input variables the interarrival time distribution $\vec{a}$ and the service time distribution $\vec{b}$ of the system are known. The random variables of the interarrival time A and service time B can only be the integer multiple of the value t_{inc}. The probability for an interarrival time of n is defined with $P(A = n) = a_n$ and the probability for a service time of m with $P(B = m) = b_m$.

Typically the interarrival time and the service time are generally distributed. In this case the single station queueing system is called G|G|1 system according to the Kendall notation (Bolch et al. 2006). Grassmann and Jain (1989) present a fast numerical method for the calculation of the waiting time distribution which is based on a Wiener-Hopf factorization of the underlying random walk. Furmans and Zillus (1996) determine the distribution of the number of customers at the arrival instant in a G|G|1-queue.

For independent service times the sojourn time distribution can be determined from the convolution of the waiting time distribution and the service time distribution (Furmans 2004). An approach for the calculation of the interdeparture time distribution comes from Jain and Grassmann (1988).

For queueing systems with multiple servers and generally distributed arrival and service times, Matzka (2011) provides an algorithm to compute the exact distribution of the number of customers at the time of arrival and waiting time. Batch arrival and service processes have been investigated by Schleyer and Furmans (2007) and a method has been developed that leads to the waiting time distribution of the G|G|1 batch arrivals queueing system. The distribution of the interdeparture time and the number of customers at the time of arrival for a batch arrivals and batch service queueing system can be determined by the methods presented by Schleyer (2007).

The decomposition approach is often used to model networks with several connected queueing systems. In the continuous time domain with generally distributed interarrival and service times the Queueing Network Analyser of Whitt (1983) is an often used approach. To analyse networks with discrete time queueing models, Furmans (2004) proposes an approach which transfers the idea of Whitt (1983) to the discrete time domain. As with Whitt (1983), the approach is based on the analysis of the individual stochastically independent G|G|1 queueing systems. The method of Grassmann and Jain (1989) and Furmans and Zillus (1996) is used to calculate the performance parameters of the individual queueing systems. The interdeparture time distribution is used to connect two queueing systems, where the interdeparture time distribution of the first queueing system corresponds to the interarrival time distribution of the second system. Furthermore, Furmans (2004) presents models for the stochastic split and merge of customer flows. The decomposition approach assumes statistical independence of the operating systems in the network. This only applies to exponentially distributed times. Otherwise a decomposition error occurs.

The described methods and models in discrete time are based on the assumption of the queueing discipline *first come, first served.* With the aim of increasing the level of detail in the modelling of real systems this assumption will be considerably extended in this work.

3 Literature Review

Know the rules well,
so you can break them effectively.
-Dalai Lama XIV.

In this chapter, a review is given on different service rules in various research and application areas and the corresponding queueing models that are used for modelling these service rules. In Section 3.1 a classification of the service rules based on the literature is presented for the research area of queueing theory and for the three selected application areas communication and computer systems, production systems and material handling systems. A literature review of queueing models with different service rules is given in Section 3.2. The models are divided into queueing models and polling models. In the chapter conclusion in Section 3.3 the research gap is defined.

3.1 Service Rules in Various Research and Application Areas

Service rules are always necessary if several customers want to be served by one resource. The rule determines which customer is served next. Service rules are used in many different application areas. In communication systems, for example, messages are sent to a receiver. For a collision-free transmission of the data packages, the processing of the packages must be clearly defined by a rule. In production systems, a dispatching policy is used to determine which job is processed next on a single machine. Similarly, for material handling systems such as a 4-way crossing, handling strategies will be used to control the service, batches and priorities of the arriving units. In addition to the three main areas of application mentioned above, there are also other applications for service rules. In Health care, service rules can be used to decide which patient in the emergency room will be operated next. Other areas of application are for example elevators and maintenance of machines.

In addition to the areas of application, different service rules have been defined in the research area of queueing theory. In Section 3.1.1 the service rules in queueing theory depicted in the literature are presented. Subsequently, the three main application areas of service rules are considered. In Section 3.1.2 the service rules in communication and computer systems on which the models of the polling systems are based are presented. The service rules in production systems from the literature are classified in Section 3.1.3 and the service rules in material handling systems are described in Section 3.1.4.

3.1.1 Service Rules in Queueing Theory

In queueing theory, service rules are called queueing disciplines or priority rules. Models in which customers are divided into different priority classes are called priority queueing models. The consideration of different queueing disciplines is an established part of the queueing theory. Authors that provide an overview of queueing disciplines in queueing theory, are for example Bolch et al. (2006), Jaiswal (1968), Kleinrock (1976), Shortle et al. (2018) and Wolff (1989).

According to Kleinrock (1976), the decision of what to serve next can be related to the relative arrival times, the service times or some function of group membership. Often only the third case is considered as a priority discipline. In the following, a more extensive definition is used, in which all three cases are included. These three cases can be used to classify the queueing disciplines according to Figure 3.1.

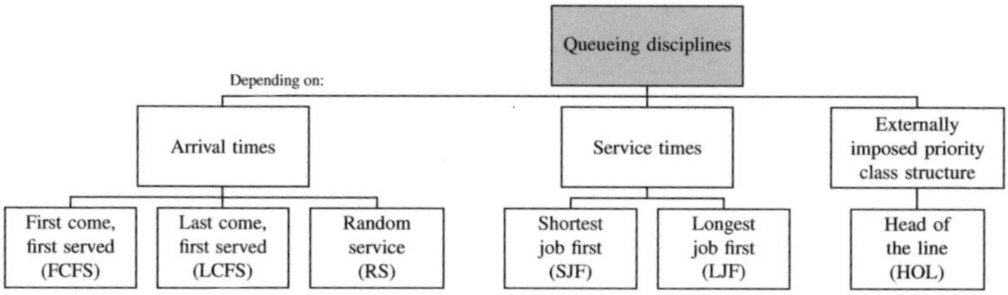

Figure 3.1: Classification of queueing disciplines in the area of queueing theory

First come, first served (FCFS) is one of the simplest queueing disciplines. Customers are served in the order of their arrival. *Last come, first served (LCFS)*, on the other hand, selects the last customer who has arrived. In the queueing discipline *random service (RS)*, the customer to be served next is selected at random. Prioritization based on the service time can be done according to the rules *shortest job first (SJF)* or *longest job first (LJF)* in which the customer is served next with the shortest/longest service time. In the *head of the line (HOL)* discipline, fixed priority classes are defined depending on external properties. The service within a class is performed according to *first-in-first-out (FIFO)*. The name of the rule HOL is derived from a model in which a separate queue is available for each priority class in front of the server and the customer of the highest priority class, which is located at the head of the line (queue), is served. Customers with lower priorities are only served if the queues of the customers with higher priorities are empty.

A further distinction in queueing discipline can be made with regard to the interruption of a service. In a preemptive queueing discipline, the service of a customer is interrupted immediately when a customer with higher priority arrives. For example, in the preemptive queueing discipline LCFS a newly arriving customer interrupts the customer currently being processed and replaces it on the server. In a non-preemptive discipline, on the other hand, the service of a customer is always continued until completion. Interruption of processing is not possible. According to Jaiswal (1968), the preemptive discipline can be further broken down into the categories resume, repeat-identical and repeat-difference. Since this work considers only non-preemptive systems, this is not discussed in detail.

3.1.2 Service Rules in Communication and Computer Systems

In communication and computer systems service rules are used in situations where several types of messages compete for access to a common resource which is available to only one type of message at a time. For example, for a multiplexer or a router with several inputs, the question arises which of the incoming messages or packets should be served next. This problem can be modelled and analysed using so-called polling models. A typical polling system consists of a number of queues, attended by a single server. A distinction is made between service routing mechanisms and service disciplines. A service routing mechanism determines the order in which the server visits the queues. The service discipline specifies the number of messages that is served during one visit of the server to a queue. In the past various survey papers dedicated to polling systems and their service routing mechanisms and service disciplines have appeared in the literature, such as Boon et al. (2011), Boxma (1991), Killat (2011), Levy and Sidi (1990) and Takagi (2000).

According to Boon et al. (2011), the server routing mechanisms can be divided into *cyclic polling*, *random polling* and *dynamic polling*. Takagi and Kleinrock (1984) names three types of service disciplines: *exhaustive service*, *gated service* and *limited service*. The class of *limited service* can be further subdivided into *customer-limited service* and *time-limited service*. Since the choice of the server routing mechanism and the service discipline are independent of each other, they can be combined without restrictions. Figure 3.2 shows the classification of the server routing mechanisms and service disciplines based on Takagi and Kleinrock (1984) and Boon et al. (2011).

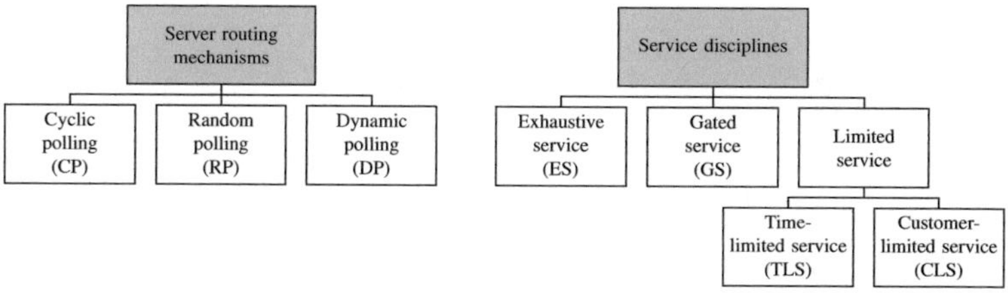

Figure 3.2: Classification of service routing mechanisms and service disciplines in the area of communication and computer systems

The traditional routing mechanism is the *cyclic polling (CL)*. In this case, the server visits the queues in a cyclic order. With a *random polling (RP)*, after leaving a queue i, the server moves to a queue j with a certain probability p_j or p_{ij} (Kleinrock and Levy 1988). In *dynamic polling (DP)*, the server's decision about the order in which queues are visited depends on the current state of the system, such as the length of the queues.

In the *exhaustive service (ES)* the server operates the customers of each queue until it is empty. Messages arriving at a queue from which customers are currently being served are also served in the same time period. In the *gated service (GS)* the server serves a queue for only those messages which are in the queue when it is polled. The messages which arrive during the service time are set aside to be served at the next service period. In the *customer-limited service (CLS)* system messages of one queue are served until either the queue is emptied or a predefined number of messages are served, whichever occurs first. The *time-limited service (TLS)* discipline behaves in the same way whereby a specified cumulated service time has expired instead of the number of messages.

An extension of *cyclic polling* is *periodic polling (PP)*, where the server visits the queue periodically according to a service order table (Baker and Rubin 1987). A special dynamic routing mechanism is the semi-dynamic server routing, in which the server decides at the end of each tour in which order it visits the queues in the next tour based on information about the queue lengths (Browne and Yechiali 1991). Furthermore, numerous hybrid variants of service disciplines can be conceived by combining the above-mentioned service disciplines. Combined service disciplines studied in the literature include *probabilistically-limited service*, *binomial-gated service*, *fractional-exhaustive service*, *mixed gated/exhaustive service* and *time-limited service disciplines with exponential time limits* (Boon et al. 2011).

3.1.3 Service Rules in Production Systems

In the production environment, service rules are called dispatching policies. The term dispatching describes the allocation of an order to a machine. A dispatching policy (also called dispatching rule) is used to select the next job to be processed from a set of jobs awaiting service. Dispatching policies have been studied extensively in single server environments. An overview of various dispatching policies is provided by Blackstone et al. (1982), Günther and Tempelmeier (2005), Lödding (2005), Nahmias (2005), Pinedo (2009) and Vollmann et al. (2004).

Dispatching rules can be classified in various ways. Lödding (2005) classifies the dispatching policies based on the optimization target into dispatching policies to increase delivery reliability, dispatching policies to increase service level and dispatching policies to increase performance. Separately, he analyses the dispatching policy shortest processing time. Blackstone et al. (1982) differentiate between rules involving processing time, rules involving due dates and rules involving shop/job characteristics other than processing time or due date. On this basis, a classification of dispatching policies can be derived according to Figure 3.3.

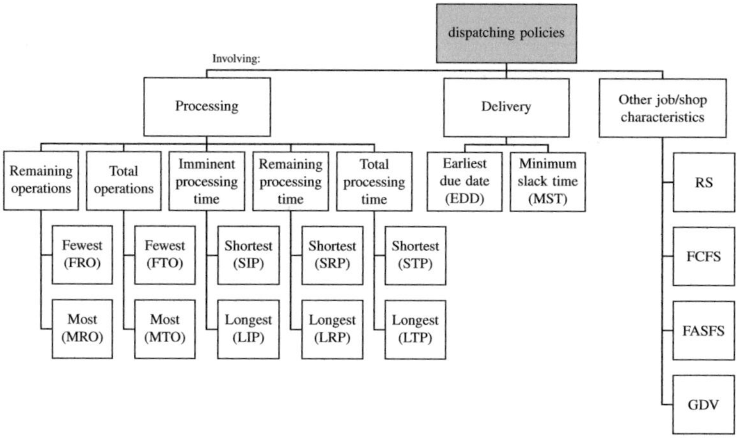

Figure 3.3: Classification of dispatching policies in the area of production

With regard to processing, the number of operations and the processing time can be taken into account. A distinction can be made between remaining operations and total operations. In addition to the remaining and total processing time, the imminent processing time can also be used as a decision criterion. The decision can be made on the fewest or most value of the criterion. Regarding the delivery there are the dispatching policies *earliest due date (EDD)* and *minimum slack time (MST)*. The slack time is the time remaining until the due date of the job that is not required for processing (slack time = due date – present date – $\sum$processing times).

The dispatching policies *random service (RS)*, *first come, first served (FCFS)*, *first arrived at shop, first served (FASFS)* and *greatest dollar value (GDV)* belong to the category of other job/shop characteristics.

In addition to the basic dispatching policies, combinations are often created. Such combinations are called composite dispatching rules. An example is the rule *apparent tardiness cost* which combines the *shortest processing time* with the *minimum slack first* policy (Pinedo 2009). Furthermore, there are heuristics that can be used to solve the assignment problem. In general a job shop with *n* jobs and *m* machines is considered. The algorithms are known as scheduling algorithms. Since this work refers to a single system, they are not considered in detail.

3.1.4 Service Rules in Material Handling Systems

In the area of material handling systems, service rules are called handling strategies. They are used to control service, batches and priorities of the arriving units, for example in conveyor systems. Arnold and Furmans (2009), Großeschallau (1984), Gudehus (1977) and Gudehus and Kotzab (2012) give a comprehensive overview of handling strategies in material handling systems.

According to Großeschallau (1984), a distinction can be made between handling of equal flows and handling of priority flows. Gudehus (1977) mentions four basic possibilities of handling strategies: *stochastic handling*, *cyclic handling*, *relative priority* and *absolute priority*. On the basis of this literature, a classification of the handling strategies is made. *Cyclic handling* is assigned to the generic term *batch handling*. The handling strategies *relative priority* and *absolute priority* are subsumed in the class *priority rules*. Figure 3.4 shows this classification of the handling strategies.

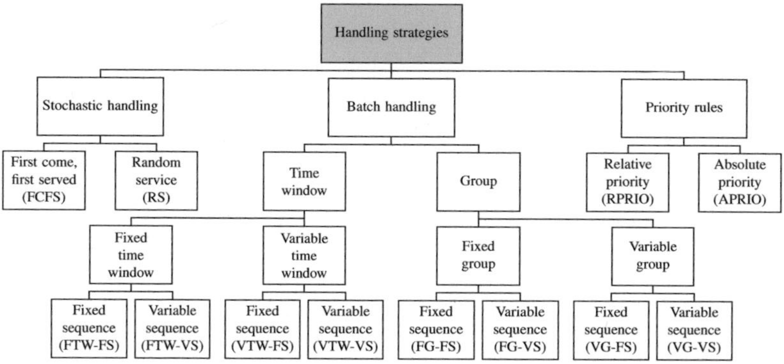

Figure 3.4: Classification of operating strategies in the area of material handling

In the classification, stochastic handling includes the handling strategies *first come, first served (FCFS)* and *random service (RS)*. In batch handling, a distinction is made between whether the batch is determined on the basis of a time window or a group number. Additionally it can be distinguished whether the time window or the group is fixed or variable depending on further parameters. A fixed or variable sequence can apply to the order in which the batches are conveyed. As an example for a *batch handling with a variable group number and variable sequence (VG-VS)*, Großeschallau (1984) mentions the batch handling with queue monitoring. With this handling strategy, there is a maximum allowed queue length for each direction. The direction whose queue length reaches the maximum allowed value next is handled until the queue is completely emptied. The switch-over takes place in the direction of the queue which has built up the maximum queue length in the meantime. The priority rules distinguish between relative and absolute priority. With *relative priority (RPRIO)*, conveying units of the lower prioritized direction are allowed to pass through the conflict area if no conveying units are waiting at the prioritized direction. *Absolute priority (APRIO)* means that conveying units of the lower prioritized direction are not allowed to interfere conveying units of the prioritized direction.

For the analysis of handling strategies Gudehus (1976) introduces the irreducible transport node of order $n+m$, which consists of n incoming transport streams and m outgoing transport streams, whereby all streams pass through a common conflict area. Since the models and analyses are limited to maximal throughput analyses of the transport node for the different handling strategies, they are not discussed further.

3.2 Analytical Models with Different Service Rules

In this section a review on analytical models from literature is given, which are used for modelling service rules. First, in Section 3.2.1, different queueing models are presented, focusing on priority queueing models and vacation models. In Section 3.2.2 an overview of polling models from the literature is given.

Since the goal of this work is to develop a discrete time model, the focus is on discrete time queueing models and discrete time polling models. Since the separation between queueing model and polling model cannot always be done unambiguously, models with one queue are assigned to the queueing models and models with several queues to the polling models.

3.2.1 Queueing Models

The term queueing theory was established by A. K. Erlang in 1908 (Leibowitz 1968). He studied problems of congestion in the telephone service for a telephone company. Since then, research has focused on describing queueing system performance from a wide variety of application areas using queueing theory. After Erlang laid the foundation for the application of queueing theory in communication systems, several scientists were interested in the problem of congestion and developed general models that can be used in more complex situations.

An important part of queueing theory are priority queueing models, which are models with different priority classes. Jaiswal (1968), Kleinrock (1976) and Wolff (1989) give a comprehensive overview about the classical priority queueing models. Vacation models are a subset of queueing models. In a vacation model, the server of a queueing system is only available part of the time. At other times it is busy serving other stations or just not available. Alfa (2010) and Doshi (1986) give an overview of queueing systems with server vacation. A structured summary and classification of the models considered in the following is given in Table 3.1.

Kleinrock (1965) proves that the mean waiting time is constant for any queue discipline and any given arrival and service time distribution under the assumption that

- all customers remain in the system customer completely serviced,
- there is a single server which is always busy if there are any customers in the system,
- preemption is allowed only if the service time distributions are exponential, and the preemption is of the preemptive resume type and
- arrival statistics are all Poisson; service statistics are arbitrary; and arrival and service statistics are all independent of each other.

This knowledge is called Kleinrock's conservation law. It can be shown that the conservation law also applies to the more general case of a G|G|1 queueing system, if the expected value of the waiting time and the service time are independent of each other (Kleinrock 1976). If Kleinrock's conservation law applies, no reduction of the total average waiting time can be achieved by prioritization. The waiting time is only shifted between the classes. Many models are developed on this basis.

Publication	Model		Characteristics			Performance measures			
	Time	Method	Arrival process	Service process	Service rule	Queue length	Waiting time	Sojourn time	Interdeparture time
Kleinrock (1965)	cont.	exact	M	G	HOL	-	$E(x)$	-	-
Kella and Yechiali (1985), Kella and Yechiali (1988)	cont.	exact	M	G	HOL	-	$E(x)$, $Var(x)$	-	-
Shanthikumar (1984), Shanthikumar (1989)	cont.	exact	M	G	FCFS, HOL, SJF	-	$E(x)$	-	-
Heidemann (1994), Heidemann and Wegmann (1997)	disc.	exact	M	G	FTW-FS, HOL	$F(x)$, $E(x)$	-	$F(x)$, $E(x)$	-
Takagi and Leung (1994)	disc.	exact	G	G	TLS	$E(x)$	$E(x)$	-	-
Alfa (1998), Alfa (2003)	disc.	exact	G	G	ES, GS, TLS, CLS	$F(x)$	$F(x)$	-	-
Walraevens et al. (2002), Walraevens et al. (2004)	disc.	exact	G	G	HOL	-	-	$F(x)$, $E(x)$	-
Lee (2001), Lee et al. (2003)	disc.	exact	G	G	HOL	-	-	$F(x)$, $E(x)$	-
Derbala (2005)	disc.	exact	M	G	HOL	-	$E(x)$	-	-
Chang and Choi (2005)	disc.	exact	G	G	ES	$F(x)$	$F(x)$	-	-
Jolai et al. (2010)	disc.	approx.	G	G	HOL	$E(x)$	$E(x)$	-	-
Zimmermann et al. (2018)	disc.	approx.	G	G	HOL	$F(x)$	$F(x)$	-	$F(x)$

Table 3.1: Literature review on queueing models

Kella and Yechiali (1985) present a methodology for the study of waiting times in a M|G|1 queueing system with several classes of customers and with single or multiple server vacations. They calculate the Laplace-Stieltjes transform (LST) and the first two moments of the waiting time of a class-k customer. The methodology is based on the observation that each model may be viewed as a special version of the basic single-class M|G|1 queueing system with server vacations. Kella and Yechiali (1988) extend this model from a non-preemptive priority queue to a preemptive model.

Shanthikumar (1989) presents a conservation identity for M|G|1 priority queues with server vacations. With a simpler approach using a level crossing analysis he confirms the results of Kella and Yechiali (1988). The presented conservation identity states that the ratios of mean waiting times in an M|G|1 queueing system with server vacation are independent of the service discipline for *first come, first served* and *shortest processing time* as well as non-preemptive priority service disciplines.

In the discrete time domain, most models refer to a specific use cases. Heidemann (1994) determines the waiting time distribution at an arrival instant of a vehicle and the delay distribution at a traffic signal. Assuming Poisson arrivals, an intersection with fixed-time-control and one lane is considered. The service rule can be described according to Section 3.1.4 as *fixed time window with fixed sequence (FTW-FS)*. For an unsignalized intersection Heidemann and Wegmann (1997) calculate the same performance parameters based on the same approach.

Takagi and Leung (1994) analyse a discrete time, single server vacation queueing system in which the length of each service period is limited by a time limit. A finite set of linear equations are solved using discrete Fourier transformation. They calculate the exact mean waiting time.

Another queueing model with server vacation in discrete time is presented by Alfa (1998). He models a *gated time-limited service*, where the server goes on a vacation as soon as all the customers within the gate have been served or the time limit has been reached (whichever occurs first). He calculates the mean number of customers waiting in the system at an arbitrary time, the amount of work in the system and the waiting time distribution using an absorbing Markov chain approach. On this basis Alfa (2003) considers a class of discrete time vacation models and presents a unified framework for analysing this class of problems. Thereby he considers the service disciplines *exhaustive service*, *gated service*, *time-limited service* and *customer-limited service*.

Walraevens et al. (2002) analyse a high- and low-priority packet delay in a queueing system with HOL priority scheduling. A Markov chain is used to determine the distributions and the expected values of the total time period a packet spends in the system for each priority class. As an application, the performance of a $N \times N$ router with output queues and two types of traffic is studied.

Lee et al. (2003) consider a discrete-time two-class queueing system with non-preemptive priority. Service times of messages of each priority class are independent and identically distributed according to a general discrete distribution function that may differ between two classes. Using the supplementary variable method and the generating function technique, they derive the joint system occupancy distribution at an arbitrary slot, and also compute the probability distributions for the sojourn time and the busy period.

In a production environment Derbala (2005) models a dispatching rule called *mean bounded priority with arrival pattern (MBPAP)*. It prioritizes jobs by an index computed for each job as a weighed sum of the proportion of time in which it has been processed and the proportion of time in which it has been waiting for processing. He derives equations for the mean waiting times of a class-k customer.

Chang and Choi (2005) analyse the performance of a finite buffer discrete time queue with bulk arrival, bulk service and vacations. In this model, the customers of a queue are served in batches until the queue is completely empty, then the server goes on vacation. Using an embedded Markov chain he calculates the steady state departure-epoch probabilities. Various performance measures such as the loss probability, the mean delay in the queue of a packet and the probability that the server is busy are calculated.

In addition Jolai et al. (2010) model a preemptive discrete time priority buffer system with partial buffer sharing with low-priority and high-priority customers based on a Markov chain. The high-priority customers have a preemptive priority over low-priority customers. To reduce the number of linear equations, a recursive numerical procedure is developed to find the steady state probabilities. For performance analysis, the expected value of the queue length and the waiting time for both low-priority and high-priority customers were calculated separately.

Zimmermann et al. (2018) analyse a discrete time G|G|1 queue with absolute priorities. The model is based on an embedded Markov-chain which is used to obtain the steady state probabilities. With an iterative calculation the distribution of the number of waiting customers, the waiting time distribution and the interdeparture time distribution is calculated.

3.2.2 Polling Models

A polling system consists of a number of queues served by a single server. There is a huge literature on polling systems that has evolved since the late 1950s. The term polling comes from the so-called polling data link control scheme, in which a central computer (server) queries each terminal (queue) on a communication line to determine whether it has to transmit information (customers). The addressed terminal sends information and the computer then switches to the next terminal to check whether that terminal has information to transmit. Takagi (1996) presents a fairly complete bibliography on polling models that contained over 700 publications. As research on polling models has continued in recent years, the number of contributions could now be well over 1,000. (Boon et al. 2011)

A comprehensive overview of polling models is provided by Bruneel and Kim (1993), Kleinrock and Levy (1988) and Takagi (1986). Boon et al. (2011) discuss the main application areas of polling systems with a comprehensive list of references and examine how these various applications can be represented and analysed via polling models. Takagi (2000) gives an introductory overview of the analysis results of the polling model and its applications to the performance evaluation of several communication protocols. He closes the paper with a survey of surveys in which he names various surveys, bibliographies and books.

The literature considered in the following is only a small selection with a focus on the discrete time domain and various server routing mechanisms and service disciplines. In Table 3.2 the considered papers are presented in a structured way.

Priority systems with switch-over times between different classes do not possess the work-conserving property, because the server is forced to be idle although work is present. Watson (1985) defines pseudo-conservation laws, which contain expressions for a weighted sum of the mean waiting times at the queues. He calls it pseudo-conservation laws because in models with switching times the amount of work is no longer independent of the service strategy. For the cases of *exhaustive service*, *gated service* and *customer-limited service* he defines pseudo-conservation laws. Boxma and Meister (1987) extend the formulas by mixed service strategies and generalize and unify the known pseudo-conservation laws.

Publication	Model		Characteristics					Performance measures			
	Time	Method	Arrival process	Service process	Switching process	Service routing	Service discipline	Queue length	Waiting time	Sojourn time	Interdeparture time
Watson (1985), Boxma and Meister (1987)	cont.	exact	M	G	G	CP	ES, GS, CLS	-	$E(x)$	-	-
Cooper and Murray (1969), Cooper (1970)	cont.	exact	M	G	-	CP	ES, GS	$E(x)$	$E(x)$	-	-
Boon and Adan (2009)	cont.	exact	M	G	G	CP	ES/GS	$F(x)$	$F(x)$	-	-
Boxma and Groenendijk (1988), Takahashi and Kumar (1995)	disc.	exact	G	G	G	CP	ES, GS, 1LS	-	$E(x)$	-	-
Boxma and Meister (1987)	disc.	approx.	M	G	G	CP	1LS	-	$E(x)$	-	-
Blanc (1990)	disc.	approx.	M	M	-	CP	ES/1LS	-	$F(x)$	-	-
Leung (1990), Leung (1991)	disc.	approx.	M	G	-	CP	1LS, CLS	$F(x)$	$F(x)$	-	-
Tran-Gia (1992)	disc.	approx.	G	G	G	CP	1LS	-	$E(x)$	-	-
Dittmann and Hübner (1993)	disc.	approx.	G	G	G	CP	GS/LS	-	$F(x)$	-	-
Coury and Harrison (1997)	disc.	approx.	G	G	-	CP	1LS	-	$F(x)$	$F(x)$	-
Frigui and Alfa (1999)	disc.	approx.	G	G	-	PP	TLS	-	$E(x)$	-	-
Beekhuizen and Resing (2009)	disc.	approx.	G	G	-	RP	RS	$F(x)$	$E(x)$	-	-
Rimmele et al. (2015)	disc.	approx.	G	G	G	CP	1LS	$F(x)$	$F(x)$	-	$F(x)$

Table 3.2: Literature review on polling models

Cooper and Murray (1969) study an *exhaustive service* and a *gated service* model. The queues of the polling model are served in cycle order. Based on a Poisson arrival process, the mean number of waiting customers and the mean cycle time are determined. The Laplace-Stieltjes transform of the cycle time distribution function is given, in a form suitable for numerical computation.

Boon and Adan (2009) consider a single-server polling system with switch-over times and introduce a new service discipline, *mixed gated/exhaustive service*, that can be used for queues with high and low priority customers. They determine the generating functions of the joint queue length distribution of all customers at visit beginnings and completions of each queue and the Laplace-Stieltjes transforms of the distributions of the cycle time, visit times and intervisit times. These distributions are used to determine the marginal queue length distributions and waiting time distributions of high and low priority customers in all queues. A pseudo-conservation law for the mean waiting times is presented, which shows, that the pseudo-conservation law also holds for polling systems with *mixed gated/exhaustive service* in some or all of the queues.

A discrete time pseudo-conservation law for mean waiting times in a multi-queue system with cyclic service and non-zero switching time is considered by Boxma and Groenendijk (1988). They obtain *exhaustive service*, *gated service* or *1-limited service* and calculate the expected value of the waiting time. *1-limited service (1LS)* is a special case of the *customer-limited service (CLS)* in which only one customer per queue is served in a cycle. Takahashi and Kumar (1995) extend the formulas to a priority multi-queue system.

Boxma and Meister (1987) present a waiting-time approximation for cyclic service systems with Poisson arrivals, switch-over times and *1-limited service*. Based on the pseudo-conservation law in the discrete time domain a simple approximative calculation approach is shown, which is exact for the completely symmetric case. They analyse the approximation accuracy by a comparison with a simulation model.

An iterative numerical technique for the evaluation of queue length distributions of multi-queue systems with one server and cyclic service discipline with Bernoulli schedules is derived by Blanc (1990). Jobs arrive at a queue according to a Poisson process and service times are assumed to be identically, exponentially distributed. The server visits queue j according to a Bernoulli schedule with parameter q_j, which includes *exhaustive service* ($q_j = 1$) and *1-limited service* ($q_j = 0$). Furthermore, the waiting time distribution is calculated with an approximative algorithm.

Leung (1990) develops an iterative numerical solution to the waiting time distributions for asymmetric token-passing systems with *1-limited service*. Customer service times and switching times have general distributions. With a set of embedded Markov chains the probability generating function for the marginal queue-length distribution for each queue at a service completion is obtained. By a numerical technique based on discrete Fourier Transforms the waiting time distribution for each queue is found.

An approximate discrete time polling system with finite capacity of waiting places and *1-limited service* is presented by Tran-Gia (1992). The analysis method is based on the use of efficient discrete convolution operations based on the Fast Fourier Transform. Using the equilibrium Markov-chain state probabilities obtained by the algorithm, the message blocking probability, the arbitrary-time state probabilities and the mean waiting time are derived.

Dittmann and Hübner (1993) study a cyclic service systems with *gated limited service*, where each queue has a fixed but individual limit. Considering general renewal input traffic, service time and switching time distribution, an approximate discrete time analysis for the cycle length and waiting time distribution is presented.

The waiting and sojourn time distributions in a class of multi-queue systems served in cyclic order with *1-limited service* in discrete time is approximated by Coury and Harrison (1997). Based on phase-type interarrival time distributions a Markov chain which represents the position of the server at each time interval is created. The algorithm for calculating the sojourn time probability density function is based on successive convolutions of the sequence.

A multi-queue system with *periodic polling* according to a pre-specified table and *time-limited service* is considered by Frigui and Alfa (1999). The analysis is based on a decomposition method. Each queue is considered separately as a queue with vacation. The visit period and vacation period distributions are obtained based on the properties of the discrete phase distribution. The visit period distribution is used to determine the average queue length using an iterative algorithm.

Beekhuizen and Resing (2009) derive an approximation of the marginal queue length distribution in a discrete time polling system with batch arrivals and fixed packet sizes. The polling server uses *random polling* and the Bernoulli service discipline (after service of queue i the server serves queue i again with probability q_i). The algorithm is based on a structured Markov chain, where the contents of one queue are stored in the level and truncated contents of the other queues in the phase.

Rimmele et al. (2015) present a 4-way-crossing under the dispatching policy round robin (synonym for *1-limited service*). The 4-way-crossing is modelled as a polling system in which a single server serves two queues cyclically. An iterative algorithm to approximate the steady state probabilities of the queues is proposed. Based on the steady state probabilities, the queue length distribution in random epochs, the waiting time distribution and the interdeparture time distribution are determined.

3.3 Chapter Conclusion

In this chapter, an overview of various service rules in different research and application areas and the corresponding queueing models used to model the service rules is given. Altogether 46 service rule classes are identified (6 in queueing theory, 12 in communication and computer systems, 16 in production systems, 12 in material handling systems) and an abbreviating name is defined for each class. The service rule classes of the different research and application areas can be partly combined with each other. Special extensions of the service rule classes can be found in the literature as well.

In a review different queueing models and polling models from the literature are presented. In Table 3.1 and Table 3.2 the considered literature is summarized and classified with respect to model type, system characteristics and considered performance measures. The literature presented is only a selection with a focus on the discrete time domain and different service rules.

The limitations of the existing literature are the following. In the existing literature, only one or a few service rules from the corresponding research or application area are considered. For an exact calculation of the performance parameters a Poisson arrival process is mostly assumed. Furthermore, the exact methods such as Kleinrock's conservation law or a pseudo-conversation law only calculate expected values. In order to determine distributions of performance parameters, approximations are often used. The switching time between two queues is often assumed to be negligibly small. For simplification, a symmetric system with the same interarrival and service times is usually considered. The interdeparture time distribution for different customer types to build networks using a decomposition method is only determined by Rimmele et al. (2015).

Overall, there is no analytical discrete time model of a multi-queue system

- with generally distributed interarrival and service times,
- which considers generally distributed switching times,
- whereby the time distributions can be different for each queue,
- with which different service rules are modelled,
- with which all performance parameter distributions are determined for each queue,
- and with which the interdeparture time distribution is determined for different departure streams.

In this work, the identified research gap is closed by creating a holistic classification of the service rules (Chapter 4) and developing a analytical model (Chapter 5) as well as a simulation model (Chapter 6) based on this classification. A detailed analysis and evaluation of the system and the service rules independent of the application areas allows to give holistic recommendations regarding the service rules (Chapter 7).

4 Classification of Service Rules

Most of the fundamental ideas of science are essentially simple, and may, as a rule, be expressed in a language comprehensible to everyone.
-Albert Einstein

As shown in Chapter 3.1, there are different service rules and classifications depending on the research or application area. However, there is no holistic classification of the service rules which contains all important service rules of the different areas. In this chapter, therefore a holistic classification is developed. It combines the various classifications from Chapter 3.1. It is assumed that a service rule (SR) makes a decision related to several queues in the same way as in a polling model (see Section 3.2.2). Within the queues, *first-in-first-out* applies. On the basis of the holistic classification the different service rules of the research and application areas can easily be modelled. This enables a generally valid mathematical modelling in Chapter 5. The holistic classification of the service rules is based on two questions:

1. Which queue is selected next?
2. How many customers are served from one queue?

The two questions are called rule categories in the following sections. Rule category 1 determines which queue should be selected next. If a queue is selected, it is possible to serve customers from this queue over several cycles. The number of services of customers from one queue is determined using rule category 2. It does not have to be a fixed number and can be dependent on system parameters such as the length of the selected queue or the waiting time of the customers of the other queues. A rule category (RC) consists of several rule classes that describe the criterion used to make the decision about the next queue to be selected or the number of services of customers from one queue. A rule class contains different rule types (RT).

The first step in the decision process is to check whether a customer is to be served from the same queue. This decision is based on the applied rule types from rule category 2. If the result is that a customer of the same queue will not be served, a new queue is selected based on the applied rule type from rule category 1. After a customer has been served, the system checks again whether another customer is to be served from this queue. If no rule type from rule category 2 is applied, a new queue is selected after each service.

In Section 4.1, rule category 1 and the associated rule classes are described. The various rule types are defined and described in detail. For rule category 2, the rule classes and rule types are defined in Section 4.2. In Section 4.3, the resulting service rules are derived by combining the rule types from rule category 1 and 2 and assigned to the various service rules from the research and application areas. Finally, the chapter is summarized in Section 4.4.

4.1 Rule Category 1: Next Queue to Be Selected

Based on the rule types of rule category 1 the decision is made which queue should be selected next. The rule category 1 consists of four rule classes. The decision about the next queue to be selected can be made based on the sequence, which means the order in which the queues are sorted (rule class *sequence*). The decision can also be based on a stochastic distribution over the queues (rule class *stochastic*). Alternatively, a largest or smallest value can be used to identify the next queue to be selected (rule class *largest/smallest value*). A decision by prioritizing the individual queues is also possible (rule class *priorities*). A total of ten rule types numbered 1.1-1.10 are assigned to the various rule classes. In Figure 4.1 the classification of the rule types 1.1-1.10 in rule category 1 is shown.

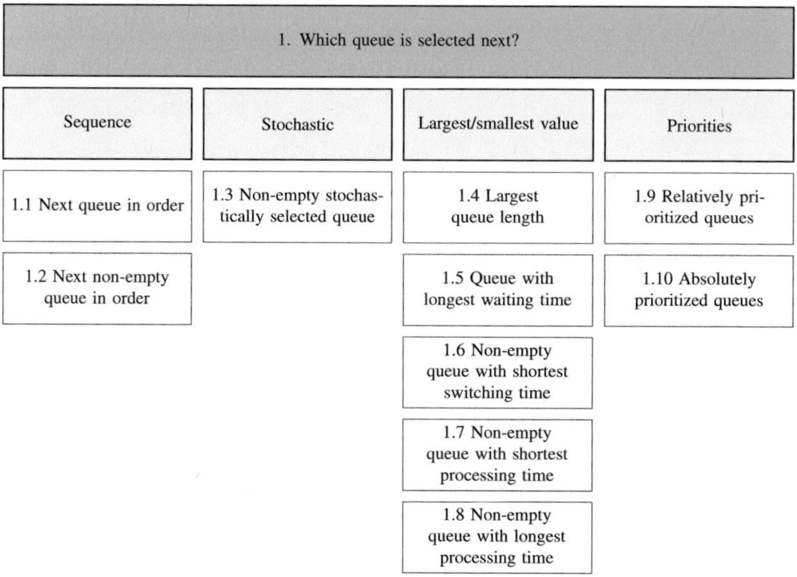

Figure 4.1: Classification of the rule types 1.1-1.10 in rule category 1

Depending on the rule type, it is possible that a decision leads to no or an ambiguous decision. For these cases additional rules have to be defined. If all queues are empty after a service, no decision can be made for rule types which are based on the criterion to select a non-empty queue. For this reason, an *additional rule for no decision* is introduced, which defines that an empty system waits until the first arrival of a customer. Afterwards, the decision is made based on the rule type. An ambiguous decision can occur when multiple queues fulfil the selection criterion. In this case, an *additional rule for ambiguous decisions* is used which defines that in case of ambiguity the queue that fulfils the selection criterion is selected, which is next in order related to the last served queue. A numbered sequence of queues is assumed. Both additional rules do not occur with all rule types. Table 4.1 summarizes the application of the additional rules.

Additional rule for...	1.1	1.2	1.3	1.4	1.5	1.6	1.7	1.8	1.9	1.10
...no decision		x	x	x	x	x	x	x	x	
...ambiguous decision				x	x	x	x	x	x	x

Table 4.1: Application of the additional rules in the different rule types of rule category 1

4.1.1 Rule Class Sequence

In rule class *sequence*, a decision is made based on the order in which the queues are sorted. Two rule types are identified which differ with regard to the handling of empty queues.

According to rule type 1.1, the next queue is selected in a fixed order. This rule is also called the *round robin* rule. After serving a customer from a queue, the next queue to be selected is the queue which is next in order related to the served queue. After the last queue of the sequence of queues is served, the system continues with the first queue. If the queue selected as the next queue is empty, it is still selected due to the fixed order in which the queues are to be served. Accordingly, the *additional rule for no decision* is not applied to this rule type. The server can therefore be empty even if customers are waiting in the not selected queues.

Similar to rule type 1.1, rule type 1.2 selects the next queue to serve based on the order of the queues. However, if the queue which is next in order related to the served queue is empty, this queue is skipped. After serving a customer of a queue, the system checks whether the following queue is not empty. If this is the case, it is selected and the first customer is served. If the queue is empty, the following queue is considered and checked again whether the queue is not empty. This process is repeated until a non-empty queue is found. For an empty system, the *additional rule for no decision* applies.

4.1.2 Rule Class Stochastic

With this rule class, the decision which queue is to be selected next is based on a stochastic distribution. For example, the distribution can be based on the arrival rates of the queues. This rule class consists of one rule type.

Rule type 1.3 makes a decision based on a given stochastic distribution. Each queue has a probability to be selected next. Based on these probabilities, the next queue is selected stochastically. Only non-empty queues are considered during the selection, so that the service can be started right after the decision. The *additional rule for no decision* applies for an empty system.

4.1.3 Rule Class Largest/Smallest Value

In rule class *largest/smallest value*, the next queue to be selected is determined by a largest or smallest value. Five rule types are assigned in this rule class, which differ with regard to the decision value. The determining values are largest queue length, shortest waiting time, shortest switching time or shortest or longest processing time. With these rule types it is possible that several queues have the same largest or smallest value at the same time and the decision based on the rule type is ambiguous. In this case, the *additional rule for ambiguous decisions* is applied, which defines that in case of ambiguity, the queue with the smallest/largest value is selected next in the order of the last served queue (see Section 4.1). If the system is empty, the *additional rule for no decision* will be used.

The decision criterion in rule type 1.4 is the queue length. The queue with the largest length, measured in customers, is selected. With rule type 1.5,the queue is selected in which the customer with the longest waiting time is held. Rule type 1.6 refers to the switching time. Here the queue with the shortest switching time to this queue is selected. The shortest or longest processing time of the first customer is used as a decision criterion in rule type 1.7 or 1.8. For stochastic switching and processing times, the decision is based on the expected value.

4.1.4 Rule Class Priorities

A decision about the next queue to be selected is made in rule class *priorities* based on a prioritization. Each queue is assigned to a specific priority. Queues with higher priority are prioritized over queues with lower priority. Two rule types are assigned to this rule class, which differ in the impact of the prioritization.

In rule type 1.9, the non-empty queue with the highest priority is selected next. If several queues have the same priority and the highest priority of the non-empty queues, the *additional rule for ambiguous decisions* applies. The *additional rule for no decision* applies for an empty system.

With *absolute prioritized queues* (rule type 1.10), a customer of a queue with higher priority must not be constrained by a customer of a queue with lower priority. This means that a customer of a higher priority queue must not wait due to a service of a customer of a lower priority queue. In order to be able to realize this rule type, additional knowledge about the next arrival of a customer in the queues is required. This is the only way to check whether a soon arriving customer in an empty higher prioritized queue has to wait because a customer of a lower prioritized queue is being served. When deciding on the next queue to be selected, the system first checks whether the highest prioritized queue is not empty. If this is the case, this queue is selected. If the highest prioritized queue is empty, the time gap until the next arrival of a customer in this queue is determined. If this time gap is too small to serve a customer from one of the lower-priority queues, the highest-priority queue is selected as the next queue. Otherwise, the queues from which a customer can be served within the time gap are determined and one of these queues is selected according to the given priority. If this queue is not empty, this queue is selected next. However, if it is empty, the time gap until the next arrival of a customer is determined for this queue and it is checked whether a customer from one of the lower priority queues can be served within this time gap. The described procedure is repeated until one queue is found as the next queue to be selected. With a completely empty system, however, the *additional rule for no decision* does not come into effect, since the additional knowledge of the arrival of the next customer can still decide which queue to serve next, so that no customer of a higher priority queue has to wait due to a customer of a lower priority queue. If several queues have the same priority and both are selected according to the procedure, the *additional rule for ambiguous decisions* applies.

4.2 Rule Category 2: Number of Customers to Be Served

If a queue is selected, it can be decided that more than one customer will be served from that queue. The number of services of customers from one queue is determined in rule category 2. The criterion used to determine the number of services is classified by three rule classes. The selection can be based on the current queue length (rule class *queue length*) or determined by a limiting value of the queue (rule class *limit value of the queue*). A limiting value related to the other queues (rule class *limit value of other queues*) can also make a decision about the repetitions of the service. Six rule types with the numbering 2.1-2.6 are assigned to the three rule classes. Figure 4.2 shows the classification for the rule category 2 with the rule types 2.1-2.6.

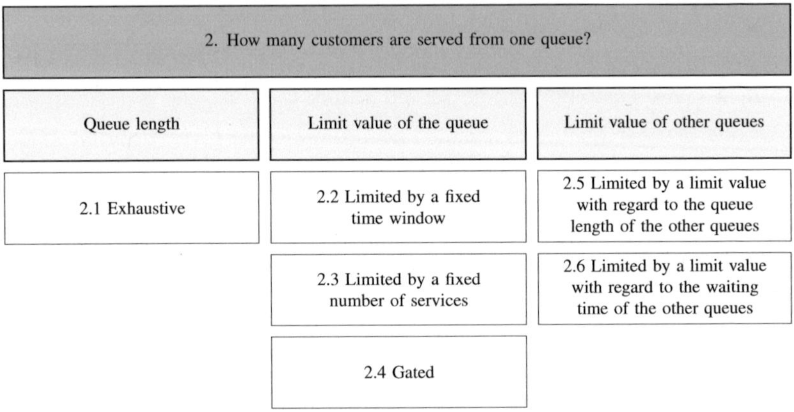

Figure 4.2: Classification of the rule types 2.1-2.6 in rule category 2

Also, in this rule category the *additional rule for no decision* applies. However, it has to be adapted for the rule types of rule category 2. The criteria that determine whether a queue will be continued can be dependent on the time. If the queue to be selected is empty after the service, the criterion to serve the same queue again can be fulfilled at first, but before a customer arrives in the system it does not apply anymore. In this case, the waiting for an arrival in the queue is cancelled when the criterion no longer applies and a new queue is determined on the basis of a rule type from rule category 1. This modified additional rule is called *additional rule for temporally changing decision*. Table 4.2 summarizes which rule types use this additional rule.

Additional rule for...	2.1	2.2	2.3	2.4	2.5	2.6
...temporally changing decisions		x			x	x

Table 4.2: Application of the additional rule in the different rule types of rule category 2

4.2.1 Rule Class Queue Length

With rule class *queue length*, the decision to serve again customers of the same queue is based on the current queue length. Accordingly, the number of services of customers of one queue is not a fixed value but depending on the current state of the system. This rule class consists of one rule type named *exhaustive*.

In rule type 2.1, customers from one queue are served until the queue is empty after service. If the queue is empty, a new queue is selected using the rule type from rule category 1. For this rule type no additional rule has to apply, because with an empty queue the criterion to serve the same queue again is not fulfilled anymore. Waiting for an arriving customer within the service of a queue cannot occur.

4.2.2 Rule Class Limit Value of the Queue

The number of services in a queue can also be defined using a limiting value related to the currently selected queue. The limiting value can be a time window, or a fixed or variable number of services.

For rule type 2.2, a fixed time window is assigned to each queue. The customers of the same queue are served until the time window has expired. If the server is in service at the end of the time window, this service is completed. Then a new queue is selected according to rule category 1. The *additional rule for temporally changing decisions* applies to this rule type. If the queue is empty after the end of the service and the time window has not expired, the system first waits for the arrival of a customer in this queue. If the time window time expires during this waiting period, the waiting for an arriving customer is cancelled and a new queue is determined by a rule class of rule category 1.

In rule type 2.3, a fixed number of services is defined for each queue. This fixed number of customers is processed one after the other. Since with this rule type the decision can be made regardless of whether a customer is in the queue or not, no additional rule has to be applied.

In the same way as rule type 2.3, in rule type 2.4 a number of services is assigned to the queue. However, the number of services is determined by the number of customers in the queue at that time when the queue is selected by a rule type of rule category 1. The number of customers already waiting in the queue at the time of selecting this queue are thus served one after the other. Customers arriving later are no longer included in the service. Therefore, this rule type is also called *gated*. As with rule type 2.1, no additional rule is required, since an empty queue and thus waiting for an arriving customer cannot occur.

4.2.3 Rule Class Limit Value of other Queues

Instead of a limiting value related to the currently selected queue, in this rule class a limiting value related to the other queues is used. Two rule types are distinguished in this rule class. The limiting value refers to the queue length or the waiting time of the other queues, respectively.

In rule type 2.5, customers from the same queue are served until the number of customers in one of the other queues reaches a fixed upper limit which is defined for each queue. As in rule type 2.2, the *additional rule for temporally changing decisions* is applied. If the limit value is not reached at any queue after the end of a customer service and the queue to be selected is empty, the system first waits for an arriving customer in the queue. If another queue reaches its limit value during this waiting period, the serving of the queue is ended and a new queue is selected.

Similar to rule type 2.5, in rule type 2.6 customers from one queue are served until the waiting time of a customer from the other queues reaches a upper limit value. Since it is assumed that *first-in-first-out* applies within the queues, the waiting time of the first customer in the queue is always the largest and reaches the limit value first. Accordingly, only the waiting time of the first customer in the queues needs to be considered. The *additional rule for temporally changing decisions* applies in the same way as for rule type 2.5.

4.3 Resulting Service Rules

A holistic classification of service rules results from Section 4.1 and 4.2. This classification is used in the following to identify the possible combinations of rule types of rule category 1 and 2 and to derive possible service rules. To establish a reference to the various research and application areas, the identified service rules are assigned to the important service rules of the various research and application areas described in Chapter 3. The corresponding tables are shown in appendix A.

4.3.1 Possible Combinations of Rule Types

To derive a service rule from the classification, a rule type of rule category 1 is chosen and if necessary combined with a rule type of rule category 2.

A service rule always consists of exactly one rule type of rule category 1. The selection of the next queue to be selected is made on the basis of one criterion. A combination of the rule types of the rule category 1 is not possible because the decisions of the different rule types could contradict each other and therefore no clear decision can be made.

The choice of a rule type of rule category 2 is optional. If no rule type of rule category 2 is selected, a new queue is selected after each service based on the rule type of rule category 1. The rule types of rule category 2 can be combined almost arbitrarily. Only rule type 2.1 and 2.4 exclude each other, because rule type 2.4 always terminates earlier than rule type 2.1. A combination therefore makes no sense, as it would lead to the same result as with a service rule where only rule type 2.4 is selected. If several rule types are combined, customers from the same queue are served until the criterion of one of the rule types is infringed. A queue is only served again if all criteria of all selected rule types of rule category 2 are fulfilled.

Rule category 1 thus results in ten possible rule types to be selected. The combination of the rule types from rule category 2 makes it possible to choose from a total of 47 possible rule type combinations. Since this rule category is optional, selecting no rule types from rule category 2 is also an option. By combining the options from rule category 1 and rule category 2, a total of 480 $(10 \cdot (47+1))$ different service rules result. The various service rules are numbered in the following and are defined as service rule 1 to service rule 480. In Table A.1 in appendix A the 480 service rules are defined by assigning the selected rule types from rule category 1 and rule category 2.

4.3.2 Assignment to the Various Research and Application Areas

To model an existing system of the various research and application areas, the resulting service rules must be applied to the rules used in the respective areas. If for example a 4-way-crossing in a conveying system is to be modelled, the handling strategies applied (see Section 3.1.4) must be modelled using the resulting service rules. Therefore, the important service rules of the various research and application areas described in Chapter 3 are assigned to the service rules from the classification.

In the area of queueing theory, most queueing disciplines can be transferred directly to the service rules of the classification. *First come, first served (FCFS)* can be represented by selecting the customer with longest waiting time (rule type 1.5). A rule type from rule category 2 is not necessary. Thus, the service rule 6 is assigned to FCFS. In the same way, the queueing disciplines *head of the line (HOL)* can be mapped directly by rule type 1.2 (*next non empty queue in order*) and thus by service rule 2. The queueing disciplines *last come, first serve (LCFS)*, on the other hand, is difficult to model due to the assumption that *first-in-first-out* applies within the queues. There is the possibility of using rule type 1.9 (*relative priority*) with queues of a capacity of one. Enough queues must be available so that one queue is always empty. The queues are assigned to different priorities and an arriving customer is assigned to an empty queue with the lowest priority, which corresponds to the highest prioritized non-empty queue. This means that the last arriving customer always has the highest priority and is served first.

Due to the fact that some rule types of the classification originate directly from the polling models, the transfer of scheduling policies of the polling systems to the identified service rules is easily possible. Since in polling models the server visits the different queues in a cyclic manner, rule type 1.1 or 1.2 can be used to map the scheduling policies of the polling models, depending on whether skipping empty queues is allowed or not. A polling systems with exhaustive service is mapped by adding rule type 2.1 to service rule 10 or 11. Equivalently, the scheduling policy *gated* is modelled using rule type 2.4 by service rule 40 or 41.

In production systems it was shown that the dispatching policy *shortest job first (SJF)* is a good rule for throughput optimization (Blackstone, Phillips and Hogg 1982). This dispatching policy can be mapped using rule type 1.7

(*Non-empty queue with the customer with shortest processing time*) to service rule 7. The dispatching policy *longest job first (SJF)* can be modelled equivalently using rule type 1.6 with service rule 6. A rule type from rule category 2 is not used. In the area of production, there are also many dispatching policies that refer to a parameter from the production network (e.g. remaining operations) or to an additional customer property (e.g. due date). These dispatching policies are mapped using rule type 1.9 (*relative priority*) in a similar way to LCFS. The number of queues is determined in relation to the additional parameter or property. For each parameter or property, the customers are assigned to the corresponding queue and prioritized according to the criterion.

The category *batch handling* in the handling strategies in the area of material handling can be mapped using rule category 2. Any handling strategy with a time window is handled according to rule type 2.2 and handling strategies with group handling according to rule types 2.1, 2.3 or 2.4. For a fixed time window with a fixed sequence, rule type 1.1 is used together with rule type 2.2 which corresponds to service rule 20. The *batch handling with queue monitoring* described by Großeschallau (1984) can be illustrated by the combination of rule type 1.4 (*largest queue length*) and rule type 2.1 (*exhaustive*) using service rule 13. The handling strategy *absolute priority* is equivalent to rule type 1.10 and thus service rule 10.

Altogether, all important service rules of the various research and application areas can be modelled using the holistic classification. The assignment of the service rules of the classification to the service rules of the various research and application areas described in Chapter 3 can be found in Table A.2 in appendix A. Using the table, the conversion of a service rule from an area can be easily realized by determining the corresponding rule types. The modelling of a real system on the basis of the model developed in Chapter 5 is thus easily possible. The unassigned service rules are also considered in the following chapters for reasons of completeness.

4.4 Chapter Conclusion

In the current chapter a holistic classification is introduced consisting of 2 rule categories, 7 rule classes and 16 rule types. The combination of the different rule types results in a total of 480 service rules that can be modelled with the classification. The service rules are defined in Table A.1 in appendix A by assigning the selected rule types from rule category 1 and rule category 2 to the service rules. Based on the resulting service rules of the classification a generic modelling of all important service rules of the various research and application areas is possible. Table A.2 shows the assignment of all service rules of the various research and application areas described in Chapter 3 to the rule types of the classification. With this table it is possible to model a service rule of an area based on the presented classification. The holistic classification can be used in Chapter 5 for the mathematical modelling of a system.

5 Modelling of a Multi-Queue System with Multiple Departure Streams under Different Service Rules

Each problem that I solved became a rule,
which served afterwards to solve other problems.
-Rene Descartes

Chapter 1 highlighted the advantage of analytical models and in this context the advantage of stochastic discrete time analysis. A Markov chain can be used to model the stochastic transition of the states (see Chapter 2). In contrast to simulation, the mathematical correlations can be mapped exactly and performance parameters can be easily calculated.

There are different models for representing material handling and production systems with different service rules. However, the models reviewed in Chapter 3 refer to special service rules and special application areas. A generally valid model is missing to model different service rules.

The classification in Chapter 4 can be used to model a wide variety of service rules from different research and application areas (see Table A.2). Based on the classification, a mathematical model will be developed in this chapter. A new model, called *multi-queue system with multiple departure streams (MQSMDS)*, is presented. It is modelled as a discrete time Markov chain.

Section 5.1 introduces the MQSMDS and basic notations of the random variables. In Section 5.2 the finite Markov chain is presented. The temporal sequence is described and the steady state and the transition probabilities are derived. Afterwards, in Section 5.3, the calculation of performance measures is shown. The chapter ends with a conclusion in Section 5.4.

5.1 System Description

A schematic representation of the *multi-queue system with multiple departure streams (MQSMDS)* is shown in Figure 5.1. It consists of N input streams and thus N queues. Each queue $i \in \{1,\ldots,N\}$ has a finite waiting room with capacity K_i. Customers who encounter a full queue upon arrival cannot enter the queue. Instead, they are rejected and lost from the system. There is no retry to enter the system. A customer will be served by the single server from queue i to sink $j \in \{1,\ldots,M\}$.

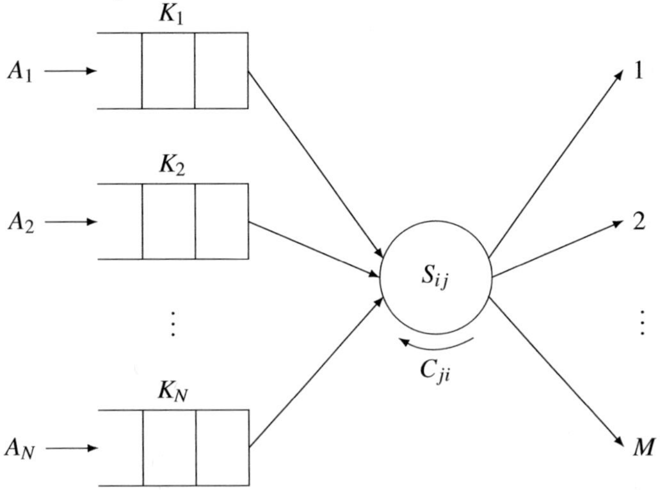

Figure 5.1: Multi-queue system with multiple departure streams (MQSMDS)

The system is observed at equally spaced time epochs $\hat{t} = 1, 2, \ldots$. Each discretization interval has a length t_{inc}. We assume that customers arrive immediately prior to the epochs and the start of a service takes place immediately after the epochs. The service consists of switching and processing. The decision about the next queue to be selected is made at the epochs. Within a time interval, switching takes place first. During the switching time, the customer to be served is still in the queue, if he has already arrived. Afterwards, the customer is processed. He leaves the queue and enters the server. The detailed temporal sequence is shown in Section 5.2.1.

The interarrival time between two consecutive arrivals at queue i is distributed according to A_i. The switching time of the server from a sink j to a queue i is distributed according to C_{ji}. The customers in queue i are processed to sink j according to the processing time S_{ij}. The transition probabilities from a queue i to the sinks are defined by $\hat{P}_i$.

The notation of the random variables whose distribution is given as input is summarized in Table 5.1. In general, $a_{i,n}$ denotes the probability $P(A_i = n)$ that the random variable A_i takes value n. Moreover, the random variables have finite lower and upper supports. Their smallest possible values are denoted by $a_{i,min}$, $c_{ji,min}$ and $s_{ij,min}$ and their largest possible values are defined by $a_{i,max}$, $c_{ji,max}$ and $s_{ij,max}$, respectively. The random variable of the transition is limited by the number of departure streams M.

	Random variable	Lower and upper support
Interarrival time of queue $i = \{1, \cdots, N\}$	A_i	$a_{i,min}, \cdots, a_{i,max}$
Switching time from sink $j = \{1, \cdots, M\}$ to queue $i = \{1, \cdots, N\}$	C_{ji}	$c_{ji,min}, \cdots, c_{ji,max}$
Processing time of queue $i = \{1, \cdots, N\}$ to sink $j = \{1, \cdots, M\}$	S_{ij}	$s_{ij,min}, \cdots, s_{ij,max}$
Transition from queue $i = \{1, \cdots, N\}$ to a sink	$\hat{P}_i$	$1, \cdots, M$

Table 5.1: Notation of the random variables whose distribution is given as input

5.2 Finite Markov Chain Model

The goal of this work is to develop a general modelling approach that allows the computation of the complete probability distributions of the MQSMDS performance measures. In order to do this, the transition matrix has to be calculated in a first step in order to be able to determine the steady state. We identify a Markov process and calculate the steady state probabilities using a discrete homogeneous finite Markov chain that is embedded at the start of service (see Section 2.2).

At first, in Section 5.2.1, a brief overview of the temporal sequence of the transition from one system state to another is given. In Section 5.2.2 it is shown how to determine the steady state probabilities based on the transition probabilities. Finally, the calculation of the transition matrix is presented in Section 5.2.3.

5.2.1 Temporal Sequence

The cycle time describes the transition time from one system state to another. It can be divided into five time intervals. Three of the intervals are possible time gaps in which the server is idle, one interval is the switching time and one is the processing time. The time gaps occur depending on the selected service rule. Figure 5.2 shows the possible time intervals of the cycle time in a temporal sequence.

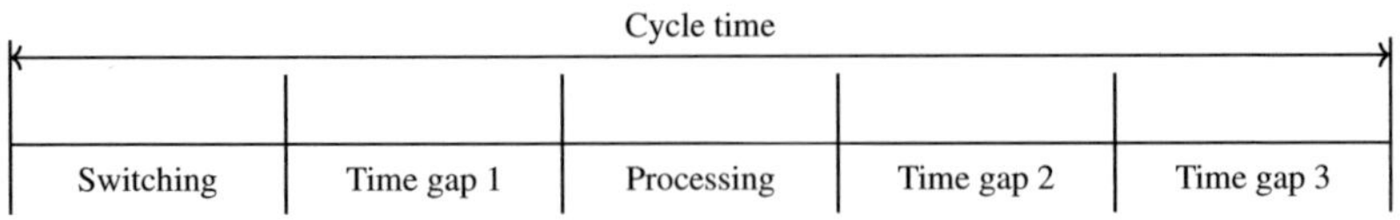

Figure 5.2: Possible time intervals of the cycle time in a temporal sequence

Time gap 1 occurs when the queue to be selected is empty after switching. This can only happen with rule type 1.1, 1.10 or 2.3. With these rule types the next queue to be selected is known, regardless of whether a customer is in this queue or not. Irrespective of whether a customer to be served is in the next queue, the server switches directly to this queue. If there is no customer in this queue, there is a time gap.

Time gap 2 only occurs with service rules of rule type 1.2, 1.3, 1.4, 1.5, 1.6, 1.7, 1.8, 1.9, 2.2, 2.5 and/or 2.6 if the system is completely empty after processing. This complies with the *additional rule for no decision* and the *additional rule for temporally changing decision*, respectively (see Section 4.1 and 4.2). To make a decision about the next queue to be selected, the server must wait until a customer arrives.

Time gap 3 appears with service rules of rule type 2.2, 2.5 and/or 2.6 if the criterion of serving customers of the same queue again is fulfilled, but there is no customer in this queue. In this case the *additional rule for temporally changing decision* applies (see Section 4.2).

Table 5.2 summarizes the occurrence of time gaps 1, 2 and 3 with the different rule types. According to the selected service rule, only time gap 1 or time gaps 2 and 3 can occur. Overall, the cycle time is calculated on the basis of the time gaps that occur together with the switching time and processing time.

Time gap	1.1	1.2	1.3	1.4	1.5	1.6	1.7	1.8	1.9	1.10	2.1	2.2	2.3	2.4	2.5	2.6
1	x									x			x			
2		x	x	x	x	x	x	x	x			x			x	x
3												x			x	x

Table 5.2: Occurrence of the time gaps with the different rule types

5.2.2 System State and Steady State Distribution

The basic system state at the time immediately before the start of the service consists of the number of customers per queue $Q_1, Q_2, \ldots, Q_N$, the residual interarrival time per queue $R_1, R_2, \ldots, R_N$, the queue of the next customer to be served Y and the sink of the last customer served Z. It can be defined as a $(2 \cdot N + 2)$-tuple as follows:

$$
\begin{aligned}
&(Q_1, Q_2, \ldots, Q_N, R_1, R_2, \ldots, R_N, Y, Z) \\
&\quad \text{with} \\
&\quad Q_i \in \{1, ..., K_i\} \quad i \in \{1, \ldots, N\} \\
&\quad R_i \in \left\{1, ..., a_{i,max}\right\} \quad i \in \{1, \ldots, N\} \\
&\quad Y \in \{1, ..., N\} \\
&\quad Z \in \{1, ..., M\}
\end{aligned}
\tag{5.1}
$$

Depending on the selected service rule or the performance parameter to be calculated, further random variables are added to the system state. For a service rule that refers to the waiting time (rule type 1.4 and 2.6) or for the calculation of the waiting time distribution (see Section 5.3.2), the waiting time of the customers at each position and queue must additionally be defined in the system state. This is necessary to model the dependency of the waiting times over the cycles. For rule type 2.2 (*limited by a fixed time window*), the system state additionally consists of the remaining time of the time window. In the same way, the system state additionally includes the remaining number of customers to be served from the same queue for a service rule with *group handling* (rule type 2.3 and 2.4). If a service rule refers to the processing time (rule type 1.6, 1.7 and 1.10), the sink of the first customer of each queue must be known in advance and is therefore additionally contained in the system state. To calculate the interdeparture time distribution (see Section 5.3.3), the last departure time per sink must be included in the system state.

	Random variable	Value at time τ	Value at time $\tau+1$	Lower and upper support
Number of customers in queue $i = \{1, \cdots, N\}$	Q_i	e_i	f_i	$0, \cdots, K_i$
Residual interarrival time of queue $i = \{1, \cdots, N\}$	R_i	g_i	h_i	$1, \cdots, a_{i,max}$
Queue of the next customer to be served	Y	k	l	$1, \cdots, N$
Sink of the last customer served	Z	u	v	$1, \cdots, M$
Waiting time of a customer at position $b = \{1, \cdots, K_i\}$ in queue $i = \{1, \cdots, N\}$	W_i^b	y_i^b	z_i^b	$0, \cdots, w_{i,max}$
Remaining time of the time window	B	o	p	$1, \cdots, TW_{max}$
Remaining number of customers	O	q	r	$1, \cdots, MN_{max}$
Sink of the first customer in queue $i = \{1, \cdots, N\}$	G_i	m_i	n_i	$1, \cdots, M$
Last departure time of sink $j = \{1, \cdots, M\}$	L_j	s_j	t_j	$d_{j,min}, \ldots, d_{j,max}$

Table 5.3: Notation of the random variables describing the system state

Table 5.3 summarizes the notation of the random variables describing the system state. Additionally the notation of the value of the random variable at a point in time τ and $\tau + 1$ is defined. The mathematical definition of the extended tuple of the system state depending on the selected service rule can be found in (B.1)-(B.5) in appendix B.1. By indexing based on the state space size x_{max} using a rectangle index function (Knott 1975), the tuple can be represented by a single system state value x.

In the steady state, a system state x is identically distributed at each observation instant (see Section 2.2). The state can be described by a random variable Λ and the steady state probability is denoted by:

$$P(\Lambda = x) = \lambda_x \tag{5.2}$$

Consider the point in time τ immediately before the start of a service. The system is in system state x. We also define a system state y in which the system is after the cycle time. The transition probability from state x to state y is denoted by p_{xy}. The transition probabilities can be calculated as shown in the following section. Knowing the transition probabilities p_{xy}, the steady state probabilities can be calculated on the basis of the following system of equations:

$$\lambda_y = \sum_x \lambda_x \cdot p_{xy} \qquad \text{with} \quad \sum_y \lambda_x = 1 \tag{5.3}$$

With this set of linear equations, the steady state distribution $\vec{\lambda}$ can be calculated. One of the equations described by (5.3) is omitted in the solution process because otherwise the set of linear equations would be overdetermined by one equation. The calculation of the transition probabilities required to determine the steady state is described in the following section.

5.2.3 Transition Probabilities

The transition probability p_{xy} is the probability of changing from a system state x to a system state y in a cycle. Mathematically, the transition probability can be defined as follows:

$$\begin{aligned} p_{xy} =& p_{(e_1,e_2,\dots,e_N,g_1,g_2,\dots,g_N,k,u),(f_1,f_2,\dots,f_N,h_1,h_2,\dots,h_N,l,v)} \\ =& P(\Lambda^{\tau+1} = y \mid \Lambda^{\tau} = x) \\ =& P(Q_1^{\tau+1} = f_1,\dots,Q_N^{\tau+1} = f_N, R_1^{\tau+1} = h_1,\dots,R_N^{\tau+1} = h_N, Y^{\tau+1} = l, \\ & Z^{\tau+1} = v \mid Q_1^{\tau} = e_1,\dots,Q_N^{\tau} = e_N, R_1^{\tau} = g_1,\dots,R_N^{\tau} = g_N, Y^{\tau} = k, \\ & Z^{\tau} = u) \end{aligned} \tag{5.4}$$

As already described in the previous section, additional random variables may be contained in the system state depending on the service rule. Accordingly, the above mathematical definition must be extended. The extended mathematical definitions can be found in the equations (B.6)-(B.10) in appendix B.2.

The calculation of the transition probabilities is divided into 4 parts. First the transition probability of queue lengths and residual interarrival time depending on a time interval is derived. In the next step the cycle time based on Section 5.2.1 is determined. Then the different service rules are modeled using the rule types from Chapter 4. Finally, the transition probability of the sink of the last customer served is determined. The transition matrix can subsequently be calculated by combining these 4 parts.

Calculation of the Transition Probability of Queue Lengths and Residual Interarrival Time Depending on a Time Interval

The transition of queue lengths and residual interarrival time can be generally described for a queue i depending on a time interval. Three cases can be distinguished, whereby the second and the third case can be further subdivided into two subcases:

- Case 1: There is no arrival in the time interval.
- Case 2: One or more arrivals take place in the time interval and the queue is not full in the succeeding system state.
  - Case 2.1: One arrival takes place in the time interval.
  - Case 2.2: Several arrivals take place in the time interval.
- Case 3: One or more arrivals take place in the time interval and the queue is full in the succeeding system state.
  - Case 3.1: Maximum one not rejected arrival takes place in the time interval.
  - Case 3.2: Several not rejected arrivals take place in the time interval.

The distinction between no arrivals, one arrival and multiple arrivals allows to mathematically model the transition of queue lengths and residual interarrival time for a time interval. Furthermore, there is a separation into no rejections possible and rejections possible. If the queue is full in the succeeding state (Case 3), rejections can taken place, but do not have to.

In Case 3.1 from n arriving customers $m \in \{n-1, n\}$ customers are rejected. In Case 3.2, at least enough customers must arrive so that the queue is full in the subsequent state. If more customers arrive, they will be rejected.

A further distinction that must be made is whether a customer is processed in the time interval or not. When processing a customer, the queue of the customer to be processed is reduced by one customer. This can be illustrated by using the following auxiliary function:

$$\Theta = \begin{cases} 1 & \text{with processing} \\ 0 & \text{without processing} \end{cases} \tag{5.5}$$

On the basis of the five cases described above and the auxiliary function Θ, the transition of queue lengths and residual interarrival time for a time interval Δ^t can be calculated as follows:

$$P_i(Q_i^{t+1} = e_i^{t+1}, R_i^{t+1} = g_i^{t+1} \mid Q_i^t = e_i^t, R_i^t = g_i^t, T^t = \Delta^t)$$
$$= \begin{cases} 1 & e_i^{t+1} = e_i^t - \Theta, \\ & g_i^t > \Delta^t, \quad g_i^{t+1} = g_i^t - \Delta^t \\ a_{i,\Delta^t - g_i^t + g_i^{t+1}} & e_i^{t+1} = e_i^t - \Theta + 1, \\ & g_i^t \le \Delta^t, \quad e_i^{t+1} < K_i \\ \sum\limits_{m=0}^{m_{max}} a_{i,\Delta^t - g_i^t - m}^{\otimes e_i^{t+1} - e_i^t - \Theta - 2} \cdot a_{i,m+g_i^{t+1}} & e_i^{t+1} > e_i^t - \Theta + 1, \\ & g_i^t \le \Delta^t, \quad e_i^{t+1} < K_i \\ a_{i,\Delta^t - g_i^t + g_i^{t+1}} + \sum\limits_{m=0}^{m_{max}} \sum\limits_{n=0}^{n_{max}} a_{i,\Delta^t - g_i^t - m}^{\otimes n} \cdot a_{i,m+g_i^{t+1}} & e_i^{t+1} \le e_i^t - \Theta + 1, \\ & g_i^t \le \Delta^t, \quad e_i^{t+1} = K_i \\ \sum\limits_{m=0}^{m_{max}} \sum\limits_{n=0}^{n_{max}} a_{i,\Delta^t - g_i^t - m}^{\otimes e_i^{t+1} - e_i^t + \Theta + n - 2} \cdot a_{i,m+g_i^{t+1}} & e_i^{t+1} > e_i^t - \Theta + 1, \\ & g_i^t \le \Delta^t, \quad e_i^{t+1} = K_i \\ 0 & \text{otherwise} \end{cases} \tag{5.6}$$

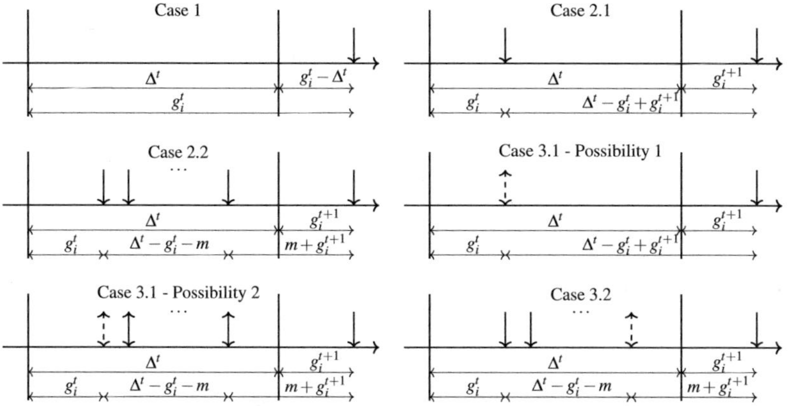

Figure 5.3: The different cases of transition of queue lengths and residual interarrival time for a time interval plotted in a time stream

The notation $a_i^{\otimes n}$ represents the n-fold convolution of the distribution a_i with itself (see Section 2.1). The various cases during a time interval of length Δ^t and the respective residual times of g_i^t and g_i^{t+1}, are depicted in Figure 5.3 using a timeline. If no customer arrives in the time interval (Case 1), the queue remains the same size (minus 1 if processing takes place). In this case the residual interarrival time must be greater than the time interval. The residual interarrival time of the succeeding state is reduced by the time interval. In Case 2, if one or more customers arrive ($g_i^t \leq \Delta^t$) and the queue is not full in the succeeding system state ($e_i^{t+1} < K_i$), the transition probability of queue lengths and residual interarrival time in a time interval is determined based on the interarrival time distribution $\vec{a}_i$. If only one customer arrives (Case 2.1), the interarrival time between this customer and the subsequent incoming customer amounts to $\Delta^t - g_i^t + g_i^{t+1}$ time units. In case that several customers arrive (Case 2.2), all combinations of arrivals that lead to a residuum of g_i^{t+1} time units in the succeeding interval have to be considered. The time between the last arrival and the end of the time interval is represented by m where m is limited by $m_{max} = \Delta^t - g_i^t - 1$.

With a full queue ($e_i^{t+1} = K_i$) (Case 3), further possible arrivals that are rejected are summed up in the transition probability. An upper boundary originates from the cycle time and the minimum interarrival time such that $n_{max} = \left\lceil \frac{\Delta^t}{a_{i,min}} \right\rceil$. In Case 3.1 it is possible that only one arrival takes place in the time interval, but also several arrivals can occur. Since both possibilities lead to the same succeeding state, the probabilities for this are added. In Case 3.2 at least $e_i^{t+1} - e_i^t + \Theta$ customers must arrive to fill the queue. In addition, n more customers may arrive who will be rejected.

If a service rule is selected which considers the waiting time, the waiting time per position and queue is part of the state space (see Section 5.2.2). In this case, due to the dependencies between waiting time, queue lengths and residual interarrival time, equation (5.6) must be extended by the transition of the waiting time of a customer W_i^b at position b in the queue i.Similar to the calculation above, three cases have to be distinguished, while Case 3 has two subcases:

- Case 1: The position b is empty at time $t+1$.
- Case 2: The position b is not empty at time t and $t+1$.
- Case 3: The position b is empty at time t and not empty at time $t+1$.
  - Case 3.1: The first arriving customer arrives at position b in the time interval.
  - Case 3.2: A customer arrives after the first arriving customer at position b in the time interval.

If position b is empty at time $t+1$ (Case 1), the waiting time in the subsequent state $y_i^{b^{t+1}}$ must be 0 at this position. However, if the position is not empty at time t and $t+1$ (Case 2), the waiting time $y_i^{b+\Theta^{t+1}}$ is calculated by $y_i^{b+\Theta^t} + \Delta^t$. Also in this case the help function Θ is used to integrate whether a customer is processed in the time interval or not. In case of processing the waiting time of position b of the succeeding state $y_i^{b^{t+1}}$ corresponds to the waiting time $y_i^{b+1^t}$ on the position $b+1$ plus the time interval Δ^t. In Case 3.1

the first arriving customer arrives at position b. Accordingly, the waiting time in the succeeding state corresponds to $\Delta^t - g_i^t$. If a customer arrives at position b after the first arriving customer (Case 3.2), the waiting time in the succeeding state depends on the interarrival time. The combination of the five cases from equation 5.6 and the four cases above would theoretically result in 20 cases. Since some combinations like 'there is no arrival in the time interval ' and 'the position b is empty at time t and not empty at time $t+1$' exclude each other, a total of 16 cases results. The mathematical calculation can be found in equation (B.11) in appendix B.3.

The transition probability of queue lengths and residual interarrival times $p_{x^{t+1}}$ for a time interval Δ^t can be determined using equation 5.6 as follows:

$$\begin{aligned} p_{x^{t+1}} =& P(\Lambda^{t+1} = x^{t+1} \mid \Lambda^t = x^t, T^t = \Delta^t) \\ =& P(Q_1^{t+1} = e_1^{t+1}, \dots, Q_N^{t+1} = e_N^{t+1}, R_1^{t+1} = g_1^{t+1}, \dots, R_N^{t+1} = g_N^{t+1} \mid \\ & Q_1^t = e_1^t, \dots, Q_N^t = e_N^t, R_1^t = g_1^t, \dots, R_N^t = g_N^t, T^t = \Delta^t) \\ =& \prod_{i=1}^{N} P_i(Q_i^{t+1} = e_i^{t+1}, R_i^{t+1} = g_i^{t+1} \mid Q_i^t = e_i^t, R_i^t = g_i^t, T^t = \Delta^t) \end{aligned} \tag{5.7}$$

The calculation of the transition probability of queue lengths, residual interarrival time and waiting time for all queues for a time interval which is required for a service rule related to the waiting time is determined in the same way using equation (B.13) in appendix B.3.

Using the equation (5.7) or (B.13), the transition probability from system state x to system state y can be calculated as a function of the cycle time Δ. The calculation is done iteratively according to the 5 time intervals of the cycle time Δ^1-Δ^5 (see Section 5.2.1). Based on the system state x, the transition probability p_{x^2} from the system state $x^1 = x$ to system state x^2 is determined dependent on Δ^1. Based on this, the transition probability p_{x^3} from the system state x^2 to system state x^3 is calculated as a function of Δ^2. This is repeated until the transition probability p_{x^6} from the system state x^5

to system state $x^6 = y$ is determined. Together with the probability to select queue l next p_{next} and the transition probability of the sink of the last customer served $p_{transition}$, which are calculated in a later step, the transition probability p_{xy} can be determined (see the final result in equation (5.41)). The transition probability from state x^3 to state x^4 is determined with $\Theta = 1$. The remaining transition probabilities are calculated with $\Theta = 0$. Figure 5.4 illustrates the sequence of the transitions and their system states in a temporal sequence.

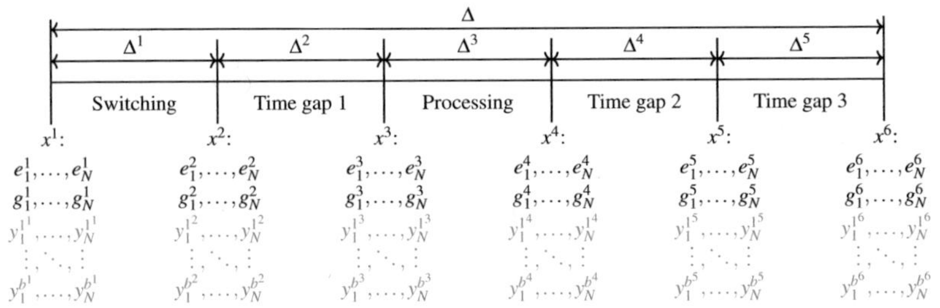

Figure 5.4: The sequence of the transitions and their system states in a temporal sequence

In order to be able to determine the transition probability from state x to y independently of the cycle time, the cycle time distribution and the distribution of its 5 parts Δ^1-Δ^5 is calculated in the next step.

Calculation of the Cycle Time Distribution

As described in Section 5.2.1 the cycle time consists of 5 time intervals whose length is named Δ^1-Δ^5 (see Figure 5.4). Δ^2, Δ^4 and Δ^5 are time gaps that can occur depending on the selected service rule. In such a time gap the server is idle and waits for a customer.

Time interval Δ^1 is the time the server takes to switch from a sink to a queue. The conditional probability p_{Δ^1} can be determined from the switching time distribution $\vec{c_{ji}}$ dependent on the sink of the last customer served u and the queue of the next customer to be served k:

$$\begin{aligned} p_{\Delta^1} &= P(T^1 = \Delta^1 \mid \Lambda^\tau = x) \\ &= P(T^1 = \Delta^1 \mid Y^\tau = k, Z^\tau = u) = c_{uk,\Delta^1} \end{aligned} \tag{5.8}$$

Time interval Δ^2 occurs when the queue to be selected is empty after switching (see Section 5.2.1). If this is the case, the length of the time gap corresponds to the residual interarrival time of the queue to be selected. The conditional probability p_{Δ^2} can thus be determined depending on the number of customers e_k^2 and the residual interarrival time g_k^2 of the queue to be selected:

$$\begin{aligned} p_{\Delta^2} &= P(T^2 = \Delta^2 \mid \Lambda^2 = x^2, \Lambda^\tau = x) \\ &= P = (T^2 = \Delta^2 \mid Q_k^2 = e_k^2, R_k^2 = g_k^2, Y^\tau = k) \\ &= \begin{cases} 1 & e_k^2 > 0, \quad \Delta^2 = 0 \\ 1 & e_k^2 = 0, \quad \Delta^2 = g_k^2 \\ 0 & \text{otherwise} \end{cases} \end{aligned} \tag{5.9}$$

Time interval Δ^3 corresponds to the processing time of a customer. Accordingly, the conditional probability p_{Δ^3} is calculated from the processing time distribution $\vec{s_{ij}}$ based on the queue of the next customer to be served k and the sink of the customer to be served v:

$$\begin{aligned} p_{\Delta^3} &= P(T^3 = \Delta^3 \mid \Lambda^\tau = x, \Lambda^{\tau+1} = y) \\ &= P(T^3 = \Delta^3 \mid Y^\tau = k, Z^{\tau+1} = v) = s_{kv,\Delta^3} \end{aligned} \tag{5.10}$$

If the system is completely empty after processing, the server must wait until a customer arrives at a queue, depending on the selected service rule. The resulting time gap is called Δ^4. As already described in 5.2.1, this time gap only occurs with rule type 1.2-1.9 and/or 2.2, 2.5 or 2.6. If a service rule is based on a rule type of rule category 2, it must first be checked after processing whether the condition of the corresponding rule type(s) of rule category 2 for processing the same queue is still fulfilled. This is determined using $p^3_{continue}$. The calculation of $p^3_{continue}$ can be found in the following section. The conditional probability of Δ^4 can be determined on the basis of five cases. If the system is not completely empty and thus the sum of the number of customers in the queues is greater than 0, $\Delta^4 = 0$ must apply (Case 1). If $p^3_{continue} = 0$, a new queue is selected and the rule type of rule category 1 RT_1 applies. If this rule type corresponds to 1.1 or 1.10, Δ^4 must also be 0 (Case 2). In the same way, with $p^3_{continue} = 1$ the time gap $\Delta^4 = 0$ if the selected service rule is not based on rule type 2.2, 2.5 and/or 2.6 (Case 3). If none of the three cases occurs, Δ^4 corresponds to the minimum residual interarrival time of the queues. Here again a case distinction regarding $p^3_{continue}$ takes place in the calculation (Cases 4 and 5). The calculation of the conditional probability p_{Δ^4} depending on the numbers of customers in the queues and the residual interarrival time of the queues can be found in the following equation:

$$
\begin{aligned}
p_{\Delta^4} &= P(T^4 = \Delta^4 \mid \Lambda^4 = x^4) \\
&= P(T^4 = \Delta^4 \mid Q^4_1 = e^4_1, \ldots, Q^4_N = e^4_N, R^4_1 = g^4_1, \ldots, R^4_N = g^4_N) \\
&= \begin{cases}
1 & \sum\limits_{n=0}^{N} e^4_n > 0, \quad \Delta^4 = 0 \\
1 & p^3_{continue} = 0, \quad RT_1 \in \{1, 10\}, \quad \Delta^4 = 0 \\
1 & p^3_{continue} = 1, \quad RT_2 \cap \{2,5,6\} = \emptyset, \quad \Delta^4 = 0 \\
1 & p^3_{continue} = 0, \quad \sum\limits_{n=0}^{N} e^4_n = 0, \quad RT_1 \notin \{1, 10\}, \\
  & \Delta^4 = \min\limits_{i=\{1,\cdots,N\}} \{g^4_i\} \\
1 & p^3_{continue} = 1, \quad \sum\limits_{n=0}^{N} e^4_n = 0, \quad RT_2 \cap \{2,5,6\} \neq \emptyset, \\
  & \Delta^4 = \min\limits_{i=\{1,\cdots,N\}} \{g^4_i\} \\
0 & \text{otherwise}
\end{cases}
\end{aligned}
\tag{5.11}
$$

The probability distribution of time interval Δ^5 is calculated using five cases. Time interval Δ^5 only occurs if $p^4_{continue} = 1$, there is no customer in this queue and the selected service rule is based on rule type 2.2, 2.5 and/or 2.6. Otherwise $\Delta^5 = 0$ applies (Case 1-3). If a customer arrives in the queue to be selected within the time in which the condition for 'continue' is fulfilled, the time gap Δ^5 corresponds to the residual interarrival g^5_k (Case 4). Otherwise Δ^5 corresponds to the time span up to the time at which the condition is no longer fulfilled (Case 5). This time span is defined by an additional interval $\ddot{\Delta}$. Depending on the number of customers e^5_k, the residual interarrival time g^5_k, the previous time intervals Δ^1-Δ^3 and the condition interval $\ddot{\Delta}$, the conditional probability p_{Δ^5} can be determined using the following equation:

$$\begin{aligned}
p_{\Delta^5} &= P(T^5 = \Delta^5 \mid \Lambda^5 = x^5, \ddot{T} = \ddot{\Delta}) \\
&= P(T^5 = \Delta^5 \mid Q^5_k = e^5_k, R^5_k = g^5_k, Y^\tau = k, \ddot{T} = \ddot{\Delta}) \\
&= \begin{cases}
1 & p^4_{continue} = 0, \quad \Delta^5 = 0 \\
1 & e^5_k > 0, \quad \Delta^5 = 0 \\
1 & RT_2 \cap \{2,5,6\} = \emptyset, \quad \Delta^5 = 0 \\
1 & p^4_{continue} = 1, \quad e^5_k = 0, \quad RT_2 \cap \{2,5,6\} \neq \emptyset, \quad g^5_k \leq \ddot{\Delta}, \\
 & \Delta^5 = g^5_k \\
1 & p^4_{continue} = 1, \quad e^5_k = 0, \quad RT_2 \cap \{2,5,6\} \neq \emptyset, \quad g^5_k > \ddot{\Delta}, \\
 & \Delta^5 = \ddot{\Delta} \\
0 & \text{otherwise}
\end{cases}
\end{aligned} \tag{5.12}$$

The probability distribution $\vec{p}_{\ddot{\Delta}}$ of the condition interval $\ddot{\Delta}$ is only calculated for service rules based on rule type 2.2, 2.5 and/or 2.6. It is determined based on the individual condition intervals $\ddot{\Delta}_2$, $\ddot{\Delta}_5$ and $\ddot{\Delta}_6$ of the rule types.

For rule type 2.2 the condition for 'continue' is fulfilled until the time window has expired. Accordingly, the condition interval for the rule type 2.2 $\ddot{\Delta}_2$

must be equal to the remaining time of the time window at the time τ minus the previous time intervals. Otherwise, $\ddot{\Delta}_2 = 0$ must apply. Mathematically, the conditional probability $p_{\ddot{\Delta}_2}$ can be described as follows:

$$\begin{aligned} p_{\ddot{\Delta}_2} &= P(\ddot{T}_2 = \ddot{\Delta}_2 \mid \Lambda^\tau = x, T^1 = \Delta^1, \ldots, T^4 = \Delta^4) \\ &= P(\ddot{T}_2 = \ddot{\Delta}_2 \mid B^\tau = o, T^1 = \Delta^1, T^2 = \Delta^2, T^3 = \Delta^3, T^4 = \Delta^4) \\ &= \begin{cases} 1 & RT_2 \cap \{2\} = \emptyset, \quad \ddot{\Delta}_2 = 0 \\ 1 & RT_2 \cap \{2\} \neq \emptyset, \quad \ddot{\Delta}_2 = o - \Delta^1 - \Delta^2 - \Delta^3 - \Delta^4 \\ 0 & \text{otherwise} \end{cases} \end{aligned} \tag{5.13}$$

With rule type 2.5, customers of one queue are served until the number of customers in another queue i reaches a limit value LV_i^1. In order to determine the condition interval $\ddot{\Delta}_5$, in a first step the condition interval $\ddot{\Delta}_{i,5}$ for the individual queues $i = \{1, 2, \cdots, N\}$ is calculated. This can be determined using four cases. If the selected service rule is not based on rule type 2.5, $\ddot{\Delta}_{i,5}$ is set to 0 (Case 1). The missing number of customers to reach the limit value can be determined by $LV_i^1 - e_i^5$. If $LV_i^1 - e_i 5 \leq 0$ the limit is already reached and $\ddot{\Delta}_{i,5} = 0$ (Case 2). If the missing number of customers is equal to 1, $\ddot{\Delta}_{i,5}$ is similar to the residual interarrival time g_i^5 (Case 3). With $LV_i^1 - e_i^5 > 1$ the probability distribution $\ddot{\Delta}_{i,5}$ is calculated based on the interarrival time distribution $\vec{a}_i$ (Case 4). This leads to the calculation:

$$\begin{aligned} &P_i(\ddot{T}_{i,5} = \ddot{\Delta}_{i,5} \mid Q_i^5 = e_i^5, R_i^5 = g_i^5) \\ &= \begin{cases} 1 & RT_2 \cap \{5\} = \emptyset, \quad \ddot{\Delta}_{i,5} = 0 \\ 1 & RT_2 \cap \{5\} \neq \emptyset, \quad LV_i^1 - e_i^5 \leq 0, \quad \ddot{\Delta}_{i,5} = 0 \\ 1 & RT_2 \cap \{5\} \neq \emptyset, \quad LV_i^1 - e_i^5 = 1, \quad \ddot{\Delta}_{i,5} = g_i^5 \\ a_{i,\ddot{\Delta}_{i,5} - g_i^5}^{\otimes LV_i^1 - e_i^5 - 2} & RT_2 \cap \{5\} \neq \emptyset, \quad LV_i^1 - e_i^5 > 1, \\ 0 & \text{otherwise} \end{cases} \end{aligned} \tag{5.14}$$

The conditional probability of $\ddot{\Delta}_5$ results from the minimum of the individual condition intervals $\ddot{\Delta}_{1,5}$-$\ddot{\Delta}_{N,5}$ with $i \neq k$ and can be calculated as follows:

$$P(\ddot{T}_5 = \ddot{\Delta}_5 \mid \ddot{T}_{1,5} = \ddot{\Delta}_{1,5}, \ldots, \ddot{T}_{N,5} = \ddot{\Delta}_{N,5}, Y^\tau = k)$$
$$= \begin{cases} 1 & RT_2 \cap \{5\} = \emptyset, \quad \ddot{\Delta}_5 = 0 \\ 1 & RT_2 \cap \{5\} \neq \emptyset, \quad \ddot{\Delta}_5 = \min\limits_{i \in (\{1,\ldots N\} \wedge i \neq k)} \{\ddot{\Delta}_{i,5}\} \\ 0 & \text{otherwise} \end{cases} \tag{5.15}$$

By summing all combinations of the individual condition intervals $\ddot{\Delta}_{1,5}$-$\ddot{\Delta}_{N,5}$ that lead to a condition interval of $\ddot{\Delta}_5$ multiplied by their probabilities, the conditional probability $p_{\ddot{\Delta}_5}$ results:

$$\begin{aligned} p_{\ddot{\Delta}_5} &= P(\ddot{T}_5 = \ddot{\Delta}_5 \mid \Lambda^5 = x^5, \Lambda^\tau = x) \\ &= P(\ddot{T}_5 = \ddot{\Delta}_5 \mid Q_1^5 = e_1^5, \ldots, Q_N^5 = e_N^5, R_1^5 = g_1^5, \ldots, R_N^5 = g_N^5, Y^\tau = k) \\ &= \sum_{\ddot{\Delta}_{1,5}=0}^{(LV_1^1 - e_1^5)\cdot a_{1,max}} \cdots \sum_{\ddot{\Delta}_{N,5}=0}^{(LV_N^1 - e_N^5)\cdot a_{N,max}} P(\ddot{T}_5 = \ddot{\Delta}_5 \mid \ddot{T}_{1,5} = \ddot{\Delta}_{1,5}, \ldots, \\ &\qquad T_{N,5} = \ddot{\Delta}_{N,5}, Y^\tau = k) \cdot \prod_{i=1}^{N} P_i(\ddot{T}_{i,5} = \ddot{\Delta}_{i,5} \mid Q_i^5 = e_i^5, R_i^5 = g_i^5) \end{aligned} \tag{5.16}$$

With a service rule based on rule type 2.6, customers of one queue are served until the waiting time of a customer in another queue i reaches a limit value of LV_i^2. As in the calculation of $\ddot{\Delta}_2$ and $\ddot{\Delta}_5$, $\ddot{\Delta}_6$ is set to 0 if the selected service rule is not based on rule type 6. Since *first-in-first-out* applies within the individual queue, the waiting time $y_i^{1^5}$ at position 1 of the queue i is the highest of this queue. When calculating the remaining time until a customer's waiting time reaches the limit value, a distinction is made between queues where at least one customer is already waiting and empty queues. If a customer is already waiting, the remaining time of this customer in queue i can be determined by $LV_i^2 - y_i^{1^5}$. If the queue i is empty, the waiting time starts at 0 as soon as the customer arrives. Accordingly, the remaining time can be calculated using $LV_i^2 + g_i^5$. The minimum of the remaining time over all queues results in the conditional interval $\ddot{\Delta}_6$, which is limited by a maximum function by the lower limit 0. The conditional probability $p_{\ddot{\Delta}_6}$ results from the following equation:

$$\begin{aligned}
p_{\ddot{\Delta}_6} =& P(\ddot{T}_6 = \ddot{\Delta}_6 \mid \Lambda^6 = x^6, \Lambda^\tau = x) = P(\ddot{T}_6 = \ddot{\Delta}_6 \mid Q_1^5 = e_1^5, \ldots, Q_N^5 = e_N^5, \\
& R_1^5 = g_1^5, \ldots, R_N^5 = g_N^5, W_1^{1^5} = y_1^{1^5}, \ldots, W_N^{1^5} = y_1^{1^5}, Y^\tau = k) \\
=& \begin{cases} 1 & RT_2 \cap \{6\} = \emptyset, \quad \ddot{\Delta}_6 = 0 \\ 1 & RT_2 \cap \{6\} \neq \emptyset, \\ & \ddot{\Delta}_6 = \max\{\min\{\min\limits_{\substack{i \in (\{1,\ldots N\} \wedge \\ i \neq k \wedge e_i^5 > 0)}} \{LV_i^2 - y_i^{1^5}\}, \min\limits_{\substack{i \in (\{1,\ldots N\} \wedge \\ i \neq k \wedge e_i^5 = 0)}} \{LV_i^2 + g_i^5\}\}, 0\} \\ 0 & \text{otherwise} \end{cases}
\end{aligned} \tag{5.17}$$

The conditional probability of $\ddot{\Delta}$ results from the minimum of the individual conditional intervals $\ddot{\Delta}_2$, $\ddot{\Delta}_5$ and $\ddot{\Delta}_6$:

$$\begin{aligned}
& P(\ddot{T} = \ddot{\Delta} \mid \ddot{T}_2 = \ddot{\Delta}_2, \ddot{T}_5 = \ddot{\Delta}_5, \ddot{T}_6 = \ddot{\Delta}_6) \\
& = \begin{cases} 1 & \ddot{\Delta} = \min\limits_{n \in (RT_2 \cap \{2,5,6\})} \{\ddot{\Delta}_n\} \\ 0 & \text{otherwise} \end{cases}
\end{aligned} \tag{5.18}$$

By summing the probabilities of all combinations of $\ddot{\Delta}_2$, $\ddot{\Delta}_5$ and $\ddot{\Delta}_6$ that lead to a condition interval of $\ddot{\Delta}$, the conditional probability $p_{\ddot{\Delta}}$ can be calculated as follows:

$$\begin{aligned}
p_{\ddot{\Delta}} &= P(\ddot{T} = \ddot{\Delta} \mid \Lambda^5 = x^5, \Lambda^\tau = x, T^1 = \Delta^1, \ldots, T^4 = \Delta^4) \\
&= \sum_{\ddot{\Delta}_2 = 0}^{TW_{max}} p_{\ddot{\Delta}_2} \cdot \sum_{\ddot{\Delta}_5 = 0}^{LV_{max}^1 \cdot a_{max}} p_{\ddot{\Delta}_5} \cdot \sum_{\ddot{\Delta}_6 = 0}^{LV_{max}^2 + a_{max}} Pp_{\ddot{\Delta}_6} \\
&\quad \cdot P(\ddot{T} = \ddot{\Delta} \mid \ddot{T}_2 = \ddot{\Delta}_2, \ddot{T}_5 = \ddot{\Delta}_5, \ddot{T}_6 = \ddot{\Delta}_6)
\end{aligned} \tag{5.19}$$

The conditional probability p_Δ of the cycle time Δ can be composed from Δ_1-Δ_5:

$$p_\Delta = P(T = \Delta \mid T^1 = \Delta^1, T^2 = \Delta^2, T^3 = \Delta^3, T^4 = \Delta^4, T^5 = \Delta^5)$$
$$= \begin{cases} 1 & \Delta = \Delta^1 + \Delta^2 + \Delta^3 + \Delta^4 + \Delta^5 \\ 0 & \text{otherwise} \end{cases} \tag{5.20}$$

So far, only the transition probability of queue lengths and residual interarrival time have been considered. In order to determine the total transition probability from system state x to system state y, in the following section the service rules based on the Rules Classes from Chapter 4 are modelled and the probability to select queue l next p_{next} is calculated.

Modelling of Service Rules

Chapter 4 presents a classification of service rules using two rule categories. A service rule consists of exactly one rule type of rule category 1 that answers the question which queue will be selected next. The probability that queue l will be selected next is defined in the following as p_{next}. In addition, the service rule can also consist of several rule types from rule category 2. This category is used to decide whether the currently selected queue should continue to be selected. In the following the probability of serving customers of the same queue again is called $p_{continue} \in \{0, 1\}$.

The probability $p_{continue}$ is determined by the product of the individual probabilities $p_{n,continue}$ of the rule types RT_2 contained in the rule category of the service rule:

$$p_{continue} = \begin{cases} 1 & \prod\limits_{n \in RT_2} p_{n,continue} = 1 \\ 0 & \prod\limits_{n \in RT_2} p_{n,continue} = 0 \end{cases} \tag{5.21}$$

If rule type 2.1 applies, the same queue will continue to be selected until it is empty (*exhaustive*). The probability $p_{1,continue}$ is 1 if the number of customers in the queue k is greater than 0:

$$p_{1,continue} = \begin{cases} 1 & e_k^t > 0 \\ 0 & \text{otherwise} \end{cases} \tag{5.22}$$

With rule type 2.2, customers of the same queue are served within a time window. This rule type is represented by an additional random variable B, which describes the remaining time of the time window (see Section 5.2.2). If the remaining time at τ is greater than the elapsed time $\Delta^1 + \Delta^2 + \ldots + \Delta^t$, the same queue will continue to be selected. Depending on the value o of the random variable B and $\Delta^1, \Delta^2, \ldots, \Delta^t$, the probability $p_{2,continue}$ results:

$$p_{2,continue} = \begin{cases} 1 & o > \sum_{n=1}^{t} \Delta^n \\ 0 & \text{otherwise} \end{cases} \tag{5.23}$$

In the same way, the probability $p_{3,continue}$ is determined for rule type 2.3 (*limited by a fixed number of services*) and rule type 2.4 (*gated*) using an additional random variable O. Since rule type 2.3 and 2.4 differ only in the upper limit of the number of customers that have to be served from the same queue, the same equation can be used for both rule types. The upper limit is later determined for each rule type using additional transition equations required for rule types with additional random variables. If the value q of the random variable O at the time τ is greater than 1, there is at least one further service of customers of the same queue. The conditional probability $p_{3,continue}$ or $p_{4,continue}$ can thus be obtained using the following equation:

$$p_{3,continue} = p_{4,continue} = \begin{cases} 1 & q > 1 \\ 0 & \text{otherwise} \end{cases} \tag{5.24}$$

For rule type 2.5, serving the same queue is limited by a limit value regard to the queue length of the other queues. If the minimum of the missing number of customers to reach the limit value $LV_i^1 - e_i^t$ across all queues is greater than 0, the limit is not yet reached at any of the other queues and the same queue will continue to be selected. So the conditional probability $p_{5,continue}$ is determined in the following way:

$$p_{5,continue} = \begin{cases} 1 & \min\limits_{i \in (\{1,\dots N\} \wedge i \neq k)} \{LV_i^1 - e_i^t\} > 0 \\ 0 & \text{otherwise} \end{cases} \tag{5.25}$$

If the service of customers of the same queue is limited by the waiting time of the other queues (rule type 2.6), the missing time until the limit value of queue i is reached $LV_i^2 - y_i^{1^t}$ must be greater than 0, so that the same queue is selected again. The probability $p_{6,continue}$ can be calculated as follows:

$$p_{6,continue} = \begin{cases} 1 & \min\limits_{i \in (\{1,\dots N\} \wedge i \neq k)} \left\{LV_i^2 - y_i^{1^t}\right\} > 0 \\ 0 & \text{otherwise} \end{cases} \tag{5.26}$$

Using the equation (5.21) with the equation (5.22)-(5.26), $p_{continue}$ can be determined at a given time $t \in \{1,\dots,5\}$ according to the 5 time intervals.

The probability of serving customers of queue l next can be determined depending on $p_{continue}$. In the case of $p_{continue} = 1$, l must correspond to the queue of the last customer served k. If $p_{continue} = 0$ then p_{next} will be determined by the individual probability $p_{RT_1,next}$ to select queue l next according to the rule category RT_1. The probability $p_{continue}$ can be obtained as follows:

$$p_{next} = \begin{cases} 1 & p_{continue} = 1, \quad l = k \\ p_{RT_1,next} & p_{continue} = 0 \\ 0 & \text{otherwise} \end{cases} \tag{5.27}$$

With rule type 1.1, the queues are selected one after the other in a fixed order. The next queue l corresponds to $k+1$ until the last queue N is reached. After queue N the service of a customer of queue 1 is started again. This relationship and thus the probability $p_{1,next}$ can be described using the following equation:

$$p_{1,next} = \begin{cases} 1 & l = k+1, \quad k < N \\ 1 & l = 1, \quad k = N \\ 0 & \text{otherwise} \end{cases} \tag{5.28}$$

While rule type 1.1 follows a fixed order, rule type 1.2 skips empty queues. For the determination of $p_{2,next}$ two cases are distinguished. If $k < l$, the sum of the number of customers in the queues $k+1$ to $l-1$ must be 0. Otherwise with $k \geq l$ the number of customers in the queues must be from $k+1$ to N and from 1 to $l-1$ in total 0. This results in the following calculation:

$$p_{2,next} = \begin{cases} 1 & f_l > 0, \quad \sum_{n=k+1}^{l-1} f_n = 0, \quad k < l \\ 1 & f_l > 0, \quad \sum_{n=k+1}^{N} f_n + \sum_{m=1}^{l-1} f_m = 0, \quad k \geq l \\ 0 & \text{otherwise} \end{cases} \tag{5.29}$$

With rule type 1.3, the queue l is selected using an additional random variable $\tilde{P}$. The probability $p_{3,next}$ is determined by the distribution $\vec{\tilde{p}}$. Since only non-empty queues are considered in this rule type, the probability $\tilde{p}_l$ is normalized by the sum of the probabilities $\tilde{p}_i$ where $i \in (\{1,...,N\} \wedge fi > 0)$ is. The probability $p_{3,next}$ can thus be described using the following equation:

$$p_{3,next} = \begin{cases} 1 & \frac{\tilde{p}_l}{\sum_{(i \in \{1,..N\} \wedge f_i > 0)} \tilde{p}_i}, \quad f_l > 0 \\ 0 & \text{otherwise} \end{cases} \tag{5.30}$$

If the decision which queue is selected next is based on a largest or smallest value (rule type 1.4-1.8), the result does not have to be unambiguous, since several queues can reach the largest/smallest value at the same time. Even with priorities (rule types 1.9-1.10), an ambiguous result is possible because several queues can have the highest priority value. If the result is not unambiguous, the queue with the corresponding largest/smallest/highest value is selected, which is next in order (*additional rule for ambiguous decision*). The set of possible queues that have taken this largest/smallest/highest value is defined as Π_n. An auxiliary function Ψ is used to check whether the next queue to be selected l complies with the rule. If the set Π_n consists only of l and possibly of k, the additional rule is fulfilled and $\Psi = 1$ (Case 1). If the set Π_n consists of at least one further queue, it must be checked whether it is in the order before l. If $k < l$ applies, no possible queue of the set $\Pi \setminus \{k,l\}$ may lie between k and l. All queues $j \in \Pi \setminus \{k,l\}$ must therefore be smaller than k (Case 2) or larger than l (Case 3). If $k > l$, all queues of the set $\Pi \setminus \{k,l\}$ must be between l and k (Case 4). The auxiliary function Ψ can be defined as follows:

$$\Psi = \begin{cases} 1 & \Pi_n \setminus \{k,l\} = \emptyset \\ 1 & j < k < l, \quad \forall j \in \Pi_n \setminus \{k,l\} \\ 1 & k < l < j, \quad \forall j \in \Pi_n \setminus \{k,l\} \\ 1 & l < j < k, \quad \forall j \in \Pi_n \setminus \{k,l\} \\ 0 & \text{otherwise} \end{cases} \qquad \forall n \in \{4,\ldots,10\} \tag{5.31}$$

The probability $p_{4,next}$-$p_{10,next}$ can be determined in the same way using the auxiliary function Ψ. If l is contained in the set Π_n of the rule class $n \in \{4,\ldots,10\}$ and the additional rule is fulfilled, then $p_{n,next} = 1$ applies:

$$p_{n,next} = \begin{cases} \Psi & l \in \Pi_n \\ 0 & \text{otherwise} \end{cases} \qquad \forall n \in \{4,\ldots,10\} \tag{5.32}$$

The set of possible queues that have taken the largest/smallest/highest value Π_n can be determined according to rule type 1.4-1.10. For rule type 1.4, the queue with the largest number of customers is selected. Rule type 1.5 determines the queue with the longest waiting time. The rule types 1.6, 1.7 and 1.8 refer to the expected value of the switching or processing time of the queues. With rule type 1.6 the queue is determined with the shortest expected value of the switching time from the sink of the last customer served. For rule type 1.7 or 1.8, the shortest or longest expected value of the processing time to the corresponding sink of the first customer in the queue is taken into account using the additional random variable O. Based on the sink of the first customer n_i in queue i with $i = \{1,2,\cdots,N\}$ of the random variable O the shortest/longest expected value of the processing time is determined. For a service rule based on relatively prioritized queues (rule type 1.9), the non-empty queue with the highest priority number PN_i is determined. For rule type 1.10, however, a customer in a prioritized queue should not have to wait for a customer in a less prioritized queue (*absolutely prioritized queues*). This means that a queue with lower priority will only be selected if the time gap until the arrival of a customer with higher priority is large enough for a customer with lower priority to be served. It is assumed that with this rule type the residual interarrival time is known in each queue. The time gap for serving an already waiting customer in a queue j can be calculated from the longest switching time $c_{vj,max}$ plus the longest processing time $s_{jn_j,max}$. If the customer of the queue j has to arrive first, the maximum of $c_{vj,max}$ and h_j is added to the longest processing time instead of the switching time. If a customer of a queue j can be processed within the residual interarrival time of an empty queue i, queue i is not the next queue to be selected. Accordingly, a queue i is only considered if the minimum of the time gaps of the other queues minus the residual interarrival time h_i is less than 0. The mathematical calculation of Π_4-Π_{10} for the rule types 1.4-1.10 is shown in the equations (5.33)-(5.39).

$$\Pi_4 = \underset{i \in \{1,...N\}}{\arg\max} \{f_i\} \tag{5.33}$$

$$\Pi_5 = \underset{i \in (\{1,...N\} \wedge f_i > 0)}{\arg\max} \left\{z_i^1\right\} \tag{5.34}$$

$$\Pi_6 = \underset{i \in (\{1,...N\} \wedge f_i > 0)}{\arg\min} \{E(C_{vi})\} \tag{5.35}$$

$$\Pi_7 = \underset{i \in (\{1,...N\} \wedge f_i > 0)}{\arg\min} \{E(S_{in_i})\} \tag{5.36}$$

$$\Pi_8 = \underset{i \in (\{1,...N\} \wedge f_i > 0)}{\arg\max} \{E(S_{in_i})\} \tag{5.37}$$

$$\Pi_9 = \underset{i \in (\{1,...N\} \wedge f_i > 0)}{\arg\max} \{PN_i\} \tag{5.38}$$

$$\Pi_{10} = \underset{\substack{i \in (\{1,...N\} \wedge (f_i > 0 \vee (\min_{j \neq i \wedge f_j > 0} \{c_{vj,max} + s_{jn_j,max} - h_i\} > 0 \wedge \\ \min_{j \neq i \wedge f_j = 0} \{\max\{c_{vj,max,h_j}\} + s_{jn_j,max} - h_i\} > 0)))}}{\arg\max} \{PN_i\} \tag{5.39}$$

As already described above, the transition probability for the service rules with additional random variables must also be determined for these variables. For example, for a service rule with a time window, the transition of the remaining time of the time window o must be calculated. In the case of $p_{continue} = 1$ the remaining time of the time window at the time $\tau + 1$ corresponds to the remaining time o at the time τ minus the cycle time Δ. If $p_{continue} = 0$, p is set to the given time window TW_l. The other transitions are determined in the same way. With rule class 2.3 and 2.4 the transition must be determined depending on whether 2.3, 2.4 or 2.3 and 2.4 apply. If rule type 2.3, 2.4 and $p_{continue} = 0$ apply, the remaining number of customers r is set to the minimum of the number of customers in the queue f_l and the fixed number of services of customers MN_l of queue l, whereby at least one customer is served for a newly selected queue. If $p_{continue} = 1$ in all three cases r is set to the remaining number of customers q at the time τ minus 1. For rule type 1.7, 1.8 and 1.10, the sink of the first customer in the queue $\hat{P}_i$ is required as an additional random variable. The sink of the first customer in the queue remains the same until the customer is served. Accordingly, the sink of the first customer n_i at τ must correspond to the sink of the first customer in the queue m_i at $\tau+1$ if $i \neq k$. However, if a customers of queue i is served, the transition probability from m_i to n_i corresponds to the given transition probability $\hat{p}_{i,n_i}$. The transition probability for all queues is derived from the product of the individual probabilities. The mathematical calculation can be found in the equations (B.14)-(B.17) in appendix B.3.

Calculation of the Transition Probability of the Sink of the Last Customer Served

The sink of the last customer served is necessary in the system state to be able to determine the corresponding switching time probability $p_{\Delta 1}$ from this sink. The conditional transition probability $p_{transition}$ of the sink of the last customer served v at the time τ is calculated in the case $RT_1 \notin \{7,8,10\}$ based on the transition distribution $\vec{p}$. If the selected service rule consists of rule type 1.7, 1.8 or 1.9, the sink of the last customer served v corresponds to the sink of the first customer in the queue m_k. This relationship is illustrated in the following equation:

$$p_{transition} = \begin{cases} 1 & RT_1 \in \{7,8,10\} \quad v = m_k \\ \hat{p}_{k,v} & RT_1 \notin \{7,8,10\} \\ 0 & \text{otherwise} \end{cases} \tag{5.40}$$

The 4 components for the calculation of the transition probability are now completely determined and will be connected in the following section.

Calculation of the Transition Probabilities

The calculation of the transition probability p_{xy} from the system state x to the system state y is done by summing all probabilities leading to y using the transition probabilities of queue lengths and residual interarrival time p_{x^2}-p_{x^6}, the conditional probabilities of the five parts of the cycle time p_{Δ^1}-p_{Δ^5}, the probability to select queue l next p_{next} and the transition probability of the sink of the last customer served $p_{transition}$:

$$\begin{aligned} p_{xy} =& P(\Lambda^{\tau+1} = y \mid \Lambda^{\tau} = x) = P(\Lambda^{\tau+1} = x^6 \mid \Lambda^{\tau} = x) \\ =& \sum_{\Delta^1=0}^{\Delta^1_{max}} p_{\Delta^1} \cdot \sum_{x^2=0}^{x_{max}} p_{x^2} \cdot \sum_{\Delta^2=0}^{\Delta^2_{max}} p_{\Delta^2} \cdot \sum_{x^3=0}^{x_{max}} p_{x^3} \cdot \sum_{\Delta^3=0}^{\Delta^3_{max}} p_{\Delta^3} \cdot \sum_{x^4=0}^{x_{max}} p_{x^4} \\ & \cdot \sum_{\Delta^4=0}^{\Delta^4_{max}} p_{\Delta^4} \cdot \sum_{x^5=0}^{x_{max}} p_{x^5} \cdot \sum_{\ddot{\Delta}=0}^{\ddot{\Delta}_{max}} p_{\ddot{\Delta}} \cdot \sum_{\Delta^5=0}^{\Delta^5_{max}} p_{\Delta^5} \cdot p_{x^6} \cdot p_{next} \cdot p_{transition} \end{aligned} \tag{5.41}$$

With an extended state space due to certain rule types in the service rule or performance parameters to be calculated, the calculation is extended by the corresponding transition probabilities. The different cases can be found in appendix B.4 in the equations (B.20)-(B.23). For a system state extended by the waiting time of the customers per position and queue, the transition probabilies $p^*_{x^2}$-$p^*_{x^6}$ of equation (B.12) are used instead of the transition probabilies p_{x^2}-p_{x^6} of equation (5.7). For the other extended system states, the additional transition probabilities $p_{1,additional}$-$p_{4,additional}$ of the equation (B.14)-(B.17) are in addition multiplied.

The steady state distribution, which can be determined from the transition matrix by solving the linear equation system (see Section 5.2.2), can be used to determine the performance parameters. The calculation of the distribution or probability of the various parameters such as queue length, waiting time, interdeparture time, sojourn time, utilization and rejection is described in the following section.

5.3 Performance Measures

The performance parameters cam be determined based on the transition probabilities and the steady state probabilities. The queue length distribution of a queue at random epochs is calculated in a similar way as the transition probabilities, where each time interval is divided into two subintervals. The waiting time distribution of a queue and the interdeparture time distribution of a sink are determined using an extended state space. From the waiting time distribution and the processing time distribution the sojourn time distribution from a queue to a sink as well as the total sojourn time distribution can be derived. A distinction can be made between server utilization and system utilization. To calculate the utilization of the server, the vacancy times of the server are considered. When determining the utilization of the system, the percentage in which the complete system (queues and the server) is not empty is used. To calculate the rejection probability of a queue, a system without capacity restrictions is considered and thus the conditional probability of rejections in a queue in a time interval is determined. The rejection probability is defined by dividing the expected value of the rejections by the expected value of the arrivals in a cycle. Table 5.4 summarizes the notation of the random variables of the performance parameters.

	Random variable	Lower and upper support
Number of customers in queue $i = \{1, \cdots, N\}$ at random epochs	$\tilde{Q}_i$	$0, \cdots, K_i$
Waiting time of a customer in queue $i = \{1, \cdots, N\}$	W_i	$w_{i,min}, \cdots, w_{i,max}$
Interdeparture time of sink $j = \{1, \cdots, M\}$	D_j	$d_{j,min} \cdots, d_{j,max}$
Sojourn time from queue $i = \{1, \cdots, N\}$ to sink $j = \{1, \cdots, M\}$	U_{ij}	$u_{ij,min} \cdots, u_{ij,max}$
Total sojourn time	U	$u_{min} \cdots, u_{max}$

Table 5.4: Notation of the random variables of the performance parameters

5.3.1 Queue Length Distributions

The distribution of the number of customers in a queue at random epochs can be calculated in the same way as the transition probability based on a temporal sequence. In addition, the time intervals are divided into two parts. Figure 5.5 shows the possible subdivisions with the corresponding states.

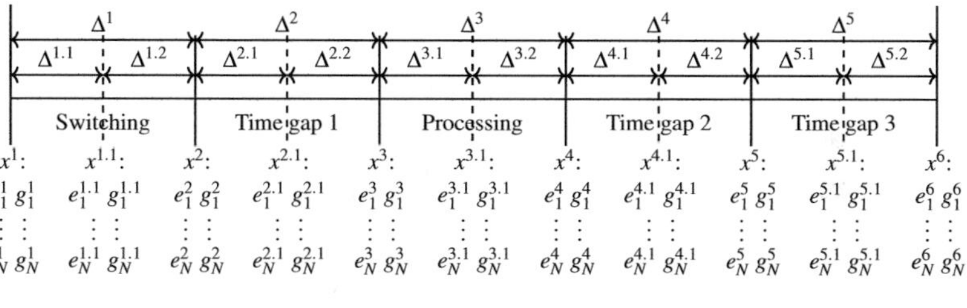

Figure 5.5: The possible subdivisions with the corresponding states in a temporal sequence

The division into two parts is necessary to consider a random point in time $\tilde{t}$. An interval is only divided if the time $\tilde{t}$ lies in the corresponding time interval. If the point in time $\tilde{t}$ is in the time interval n, the subinterval $\Delta^{n.1}$ is the time between the beginning of the time interval n and the time $\tilde{t}$. The subinterval $\Delta^{n.2}$ is the remaining time from the time $\tilde{t}$ to the end of the time interval. Otherwise, $\Delta^{n.1} = 0$ and $\Delta^{n.2} = \Delta^n$ apply. Mathematically this can be described as follows:

$$p_{\Delta^{n.1},\Delta^{n.2}} = P(T^{n.1} = \Delta^{n.1}, T^{n.2} = \Delta^{n.2} \mid T^1 = \Delta^1, \ldots, T^n = \Delta^n, \tilde{T} = \tilde{t})$$

$$= \begin{cases} 1 & \tilde{t} > \sum_{l=1}^{n-1} \Delta^l, \quad \tilde{t} \leq \sum_{l=1}^{n} \Delta^l, \quad \Delta^{n.1} = \tilde{t} - \sum_{l=1}^{n-1} \Delta^l, \\ & \Delta^{n.2} = \sum_{l=1}^{n} \Delta^l - \tilde{t} \\ 1 & \tilde{t} > \sum_{l=1}^{n} \Delta^l, \quad \Delta^{n.1} = 0, \quad \Delta^{n.1} = \Delta^n \\ 1 & \tilde{t} \leq \sum_{l=1}^{n-1} \Delta^l, \quad \Delta^{n.1} = 0, \quad \Delta^{n.1} = \Delta^n \\ 0 & \text{otherwise} \end{cases} \tag{5.42}$$

With the same approach, the number of customers in a queue at $\tilde{t}$ can be determined depending on the system state at the beginning of the time interval in which the point in time $\tilde{t}$ lies. If the time $\tilde{t}$ is in the corresponding time interval n, the queue length at $\tilde{t}$ is equal to the number of customers $e_i^{n.1}$ after the first subinterval $\Delta^{n.1}$. This results in the following calculation:

$$p_{\tilde{e}_i} = P(\tilde{Q}_i = \tilde{e}_i \mid Q_i^{1.1} = e_i^{1.1}, \ldots Q_i^{5.1} = e_i^{5.1}, T^1 = \Delta^1, \ldots, T^4 = \Delta^4, \tilde{T} = \tilde{t})$$
$$= \begin{cases} 1 & \tilde{t} \leq \Delta^1, \quad \tilde{e}_i = e_i^{1.1} \\ 1 & \tilde{t} < \Delta^1, \quad \tilde{t} \leq \Delta^1 + \Delta^2, \quad \tilde{e}_i = e_i^{1.2} \\ 1 & \tilde{t} < \Delta^1 + \Delta^2, \quad \tilde{t} \leq \Delta^1 + \Delta^2 + \Delta^3, \quad \tilde{e}_i = e_i^{1.3} \\ 1 & \tilde{t} < \Delta^1 + \Delta^2 + \Delta^3, \quad \tilde{t} \leq \Delta^1 + \Delta^2 + \Delta^3 + \Delta^4, \quad \tilde{e}_i = e_i^{1.4} \\ 1 & \tilde{t} < \Delta^1 + \Delta^2 + \Delta^3 + \Delta^4, \quad \tilde{t} \leq \Delta^1 + \Delta^2 + \Delta^3 + \Delta^4 + \Delta^5, \\ & \tilde{e}_i = e_i^{1.5} \\ 0 & \text{otherwise} \end{cases} \tag{5.43}$$

The probability to randomly observe the queue at $\tilde{t}$ in a cycle interval of length Δ is $1/\Delta$. Larger cycle intervals make up a larger share of the total time (Schwarz and Epp 2016). Thus, the probability of an observation within a cycle interval of length Δ has to be weighted with its length. The weighting is based on the expected value of the cycle time. The expected value can be determined by adding all probabilities leading to a cycle time of Δ and multiplying them by this cycle time length. Since the probability of the time interval Δ^3 and thus the following time intervals are dependent on the following system state $y = x^6$, this is again added up over all possible following system states. Together with the steady state distribution $\vec{\lambda}$, the expected value of the cycle time $E(T)$ can be determined according to the equation (5.44).

$$
\begin{aligned}
E(T) = &\sum_{\Delta=1}^{\Delta_{max}} \Delta \cdot \sum_{x=0}^{x_{max}} \lambda_x \cdot \sum_{y=x^6=0}^{x_{max}} \sum_{\Delta^1=0}^{\Delta^1_{max}} p_{\Delta^1} \cdot \sum_{x^2=0}^{x_{max}} p_{x^2} \cdot \sum_{\Delta^2=0}^{\Delta^2_{max}} p_{\Delta^2} \cdot \sum_{x^3=0}^{x_{max}} p_{x^3} \\
&\cdot \sum_{\Delta^3=0}^{\Delta^3_{max}} p_{\Delta^3} \cdot \sum_{x^4=0}^{x_{max}} p_{x^4} \cdot \sum_{\Delta^4=0}^{\Delta^4_{max}} p_{\Delta^4} \cdot \sum_{x^5=0}^{x_{max}} p_{x^5} \cdot \sum_{\ddot{\Delta}=0}^{\ddot{\Delta}_{max}} p_{\ddot{\Delta}} \cdot \sum_{\Delta^5=0}^{\Delta^5_{max}} p_{\Delta^5} \cdot p_{x^6} \\
&\cdot p_{next} \cdot p_{transition} \cdot p_{\Delta}
\end{aligned} \tag{5.44}
$$

The conditional probability of the number of customers in a queue at random epochs depending on the system state x can be calculated by summing all probabilities leading to the number of customers in a queue $\tilde{e}_i$ at time $\tilde{t}$. The probabilities are calculated and summed for all possible points in time $0, 1, \ldots, \Delta_{max} - 1$. This summed probability value is normalized by division with the expected value of the cycle time. Weighted with the steady state distribution $\vec{\lambda}$ the distribution of the number of customers $\vec{\tilde{q}}_i$ in a queue i at random epochs can be derived:

$$
\begin{aligned}
\tilde{q}_{i,\tilde{e}_i} = &P_i(\tilde{Q}_i = \tilde{e}_i) = \frac{1}{E(T)} \cdot \sum_{x=0}^{x_{max}} \lambda_x \cdot \sum_{\tilde{t}=0}^{\Delta_{max}-1} \sum_{y=x^6=0}^{x_{max}} \sum_{\Delta^1=0}^{\Delta^1_{max}} p_{\Delta^1} \\
&\cdot \sum_{\Delta^{1.1}=0}^{\Delta^1_{max}} \sum_{\Delta^{1.2}=0}^{\Delta^1_{max}} p_{\Delta^{1.1},\Delta^{1.2}} \cdot \sum_{x^{1.1}=0}^{x_{max}} p_{x^{1.1}} \cdot \sum_{x^2=0}^{x_{max}} p_{x^2} \cdot \sum_{\Delta^2=0}^{\Delta^2_{max}} p_{\Delta^2} \\
&\cdot \sum_{\Delta^{2.1}=0}^{\Delta^2_{max}} \sum_{\Delta^{2.2}=0}^{\Delta^1_{max}} p_{\Delta^{2.1},\Delta^{2.2}} \cdot \sum_{x^{2.1}=0}^{x_{max}} p_{x^{2.1}} \cdot \sum_{x^3=0}^{x_{max}} p_{x^3} \cdot \sum_{\Delta^3=0}^{\Delta^3_{max}} p_{\Delta^3} \\
&\cdot \sum_{\Delta^{3.1}=0}^{\Delta^3_{max}} \sum_{\Delta^{3.2}=0}^{\Delta^1_{max}} p_{\Delta^{3.1},\Delta^{3.2}} \cdot \sum_{x^{3.1}=0}^{x_{max}} p_{x^{3.1}} \cdot \sum_{x^4=0}^{x_{max}} p_{x^4} \cdot \sum_{\Delta^4=0}^{\Delta^4_{max}} p_{\Delta^4} \\
&\cdot \sum_{\Delta^{4.1}=0}^{\Delta^4_{max}} \sum_{\Delta^{4.2}=0}^{\Delta^4_{max}} p_{\Delta^{4.1},\Delta^{4.2}} \cdot \sum_{x^{4.1}=0}^{x_{max}} p_{x^{4.1}} \cdot \sum_{x^5=0}^{x_{max}} p_{x^5} \cdot \sum_{\ddot{\Delta}=0}^{\ddot{\Delta}_{max}} p_{\ddot{\Delta}} \cdot \sum_{\Delta^5=0}^{\Delta^5_{max}} p_{\Delta^5} \\
&\cdot \sum_{\Delta^{5.1}=0}^{\Delta^5_{max}} \sum_{\Delta^{5.2}=0}^{\Delta^5_{max}} p_{\Delta^{5.1},\Delta^{5.2}} \cdot \sum_{x^{5.1}=0}^{x_{max}} p_{x^{5.1}} \cdot p_{x^6} \cdot p_{next} \cdot p_{transition} \cdot p_{\tilde{e}_i}
\end{aligned} \tag{5.45}
$$

5.3.2 Waiting Time Distributions

The waiting time distribution of a queue is obtained by calculating the steady state with a system state extended by the waiting time of a customer W_i^b at position b in a queue i. This calculation can be performed in the same way as for a service rule based on rule type 1.5 and 2.6 (see Section 5.2.2 and 5.2.3). When calculating the waiting time of a customer, the total waiting time from arrival to the start of processing is taken into account. Accordingly, only the customer to be processed is considered in a time interval. The waiting time of the next customer to be processed depending on a system state x consists of the waiting time y_k^1 of the first customer in the queue k at the time of the system state x. The time interval Δ^1 and the time interval Δ^2 are added to this. If the customer to be processed arrives during the time interval, the residual interarrival time g_k is subtracted. The conditional probability of the waiting time depending on the system state x and time intervals Δ^1 and Δ^2 is then calculated as follows:

$$p_{i,\omega_i} = P_i(W_i = \omega_i \mid \Lambda^\tau = x, T^1 = \Delta^1, T^2 = \Delta^2)$$
$$= \begin{cases} 1 & e_k > 0, \quad \omega_k = y_k^1 + \Delta 1 + \Delta 2, \quad i = k \\ 1 & e_k = 0, \quad \omega_k = y_k^1 + \Delta 1 + \Delta 2 - g_k, \quad i = k \\ 0 & \text{otherwise} \end{cases} \tag{5.46}$$

By multiplying the conditional probability with all possible probabilities for Δ^1, x^2 and Δ^2 and weighting this with the steady state distribution, the waiting time distribution $\vec{w}_i^*$ of the queue i can be determined:

$$w_{i,\omega_i}^* = P_i(W_i^* = \omega_i) = \sum_{x=0}^{x_{max}} \lambda_x \cdot \sum_{\Delta^1=0}^{\Delta^1_{max}} p_{\Delta^1} \cdot \sum_{x^2=0}^{x_{max}} p_{x^2} \cdot \sum_{\Delta^2=0}^{\Delta^2_{max}} p_{\Delta^2} \cdot p_{i,\omega_i} \tag{5.47}$$

However, this distribution is a defective distribution since only the processed customer in the current time interval is taken into account when calculating the waiting time (Schwarz and Epp 2016). To determine the waiting time distribution $\vec{w}_i$ of a queue i the distribution $\vec{w_i^*}$ must be normalized:

$$w_{i,\omega_i} = P_i(W_i = \omega_i) = \frac{w^*_{i,\omega_i}}{\sum\limits_{n=0}^{w_{i,max}} w^*_{i,\omega_i}} \tag{5.48}$$

5.3.3 Interdeparture Time Distributions

Similar to the calculation of the waiting time distribution of a queue, the interdeparture time distribution of a sink is determined using an extended system state. The system state is extended by the last departure time L_j in a sink $j \in \{1,...,M\}$. The definition of the state space and the calculation of the transition probability can be found in appendix B in (B.5), (B.10) and (B.24).

Two cases can be distinguished when determining the conditional transition probability of the last departure time of the sink j. If the sink j is the sink in which a customer is served in the time interval, the departure time of this sink at $\tau + 1$ is $\Delta^4 + \Delta^5$. If no customer passes into the sink, the last departure time of the sink is increased by Δ. The calculation of the transition probability of the additional parameter can be found in the equations (B.18) and (B.19) in appendix B.3.

The interdeparture time of a sink can be determined from the last departure time of the sink s_j plus the time intervals Δ^1, Δ^2 and Δ^3:

$$\begin{aligned} p_{j,\delta_j} &= P_j(D_j = \delta_j \mid \Lambda^\tau = x, \Lambda^{\tau+1} = y, T^1 = \Delta^1, T^2 = \Delta^2, T^3 = \Delta^3) \\ &= \begin{cases} 1 & \delta_j = s_j + \Delta^1 + \Delta^2 + \Delta^3, \quad j = v \\ 0 & \text{otherwise} \end{cases} \end{aligned} \tag{5.49}$$

The interdeparture time distribution $\vec{d}_j^*$ can be calculated similarly to (5.47):

$$d^*_{j,\delta_j} = P_i(D^*_j = \delta_j) = \sum_{x=0}^{x_{max}} \lambda_x \cdot \sum_{y=x^6=0}^{x_{max}} \sum_{\Delta^1=0}^{\Delta^1_{max}} p_{\Delta^1} \cdot \sum_{x^2=0}^{x_{max}} p_{x^2} \cdot \sum_{\Delta^2=0}^{\Delta^2_{max}} p_{\Delta^2}$$
$$\cdot \sum_{x^3=0}^{x_{max}} p_{x^3} \cdot \sum_{\Delta^3=0}^{\Delta^3_{max}} p_{\Delta^3} \cdot \sum_{x^4=0}^{x_{max}} p_{x^4} \cdot \sum_{\Delta^4=0}^{\Delta^4_{max}} p_{\Delta^4} \cdot \sum_{x^5=0}^{x_{max}} p_{x^5} \cdot \sum_{\ddot{\Delta}=0}^{\ddot{\Delta}_{max}} p_{\ddot{\Delta}} \cdot \sum_{\Delta^5=0}^{\Delta^5_{max}} p_{\Delta^5} \tag{5.50}$$
$$\cdot p_{x^6} \cdot p_{next} \cdot p_{transition} \cdot p_{j,\delta_j}$$

This is also a defective distribution, since only the sink of the processed customer in the current time interval is taken into account. Finally, the interdeparture time distribution $\vec{d}_j$ can be calculated by normalization:

$$d_{j,\delta_j} = P_j(D_j = \delta_j) = \frac{d^*_{j,\delta_j}}{\sum_{n=0}^{d_{j,max}} d^*_{j,\delta_j}} \tag{5.51}$$

5.3.4 Sojourn Time Distributions

The sojourn time distribution $\vec{u_{ij}}$ from a queue i to a sink j can be determined from the convolution of the waiting time distribution $\vec{w}_i$ and the processing time distribution $\vec{s_{ij}}$:

$$\vec{u_{ij}} = \vec{w}_i \otimes \vec{s_{ij}} \tag{5.52}$$

The total sojourn time distribution $\vec{u}$ is calculated by weighting. The probability $u_{ij}(\vartheta)$ of a sojourn time ϑ from a queue i to a sink j is weighted on the basis of the throughput $\frac{1}{E(A_i)}$ of the queue i weighted with the inverse probability of the rejection probability $p_{i,rejection}$ (see Section 5.3.6) in relation to the total throughput and the transition probability $\hat{p}_{i,j}$:

$$u_\vartheta = P(U = \vartheta) = \sum_{i=1}^{N} \sum_{j=1}^{M} \frac{\frac{1}{E(A_i)} \cdot (1 - p_{i,rejection})}{\sum_{n=1}^{N} \frac{1}{E(A_n)} \cdot (1 - p_{n,rejection})} \cdot \hat{p}_{i,j} \cdot u_{ij}(\vartheta) \tag{5.53}$$

5.3.5 Utilization

A distinction is made between the utilization of the server and the utilization of the system. The additional consideration of the utilization of the system is necessary, since no statement can be made about whether the system is overloaded or not on the basis of the utilization of the server. Depending on the selected service rule, the server may be empty even though customers are waiting in the queue. The utilization of the server is therefore less or equal to the utilization of the system. However, the utilization of the server and the comparison of this with the utilization of the system can be used as evaluation criteria for a service rule (see Chapter 7).

In the long run, the utilization of the server can be calculated in a stable system from the relation of the time in which the server is empty to the total time in which the system is observed. In a cycle the server is empty when the time gaps 1, 2 or 3 occur (see Section 5.2.1). Accordingly, the expected values of the length of this time gaps have to be determined. Equivalent to equation (5.44) the expected values $E(T^2)$, $E(T^4)$ and $E(T^5)$ are calculated by summing all combinations of the conditional probabilities leading to the time intervals. Since the time interval Δ^2 is not dependent on a subsequent system state, the calculation of the expected value $E(T^2)$ can be done without summing all probabilities of transition to the subsequent system states:

$$E(T^2) = \sum_{\Delta^2=0}^{\Delta^2_{max}} \Delta^2 \cdot \sum_{x=0}^{x_{max}} \lambda_x \cdot \sum_{\Delta^1=0}^{\Delta^1_{max}} p_{\Delta^1} \cdot \sum_{x^2=0}^{x_{max}} p_{x^2} \cdot p_{\Delta^2} \tag{5.54}$$

$$\begin{aligned} E(T^4) = & \sum_{\Delta^4=0}^{\Delta^4_{max}} \Delta^4 \cdot \sum_{x=0}^{x_{max}} \lambda_x \cdot \sum_{y=x^6=0}^{x_{max}} \sum_{\Delta^1=0}^{\Delta^1_{max}} p_{\Delta^1} \cdot \sum_{x^2=0}^{x_{max}} p_{x^2} \cdot \sum_{\Delta^2=0}^{\Delta^2_{max}} p_{\Delta^2} \cdot \sum_{x^3=0}^{x_{max}} p_{x^3} \\ & \cdot \sum_{\Delta^3=0}^{\Delta^3_{max}} p_{\Delta^3} \cdot \sum_{x^4=0}^{x_{max}} p_{x^4} \cdot \sum_{x^5=0}^{x_{max}} p_{x^5} \cdot \sum_{\ddot{\Delta}=0}^{\ddot{\Delta}_{max}} p_{\ddot{\Delta}} \cdot \sum_{\Delta^5=0}^{\Delta^5_{max}} p_{\Delta^5} \cdot p_{x^6} \cdot p_{next} \\ & \cdot p_{transition} \cdot p_{\Delta^4} \end{aligned} \tag{5.55}$$

$$E(T^5) = \sum_{\Delta^5=0}^{\Delta^5_{max}} \Delta^5 \cdot \sum_{x=0}^{x_{max}} \lambda_x \cdot \sum_{y=x^6=0}^{x_{max}} \sum_{\Delta^1=0}^{\Delta^1_{max}} p_{\Delta^1} \cdot \sum_{x^2=0}^{x_{max}} p_{x^2} \cdot \sum_{\Delta^2=0}^{\Delta^2_{max}} p_{\Delta^2} \cdot \sum_{x^3=0}^{x_{max}} p_{x^3}$$
$$\cdot \sum_{\Delta^3=0}^{\Delta^3_{max}} p_{\Delta^3} \cdot \sum_{x^4=0}^{x_{max}} p_{x^4} \cdot \sum_{\Delta^4=0}^{\Delta^4_{max}} p_{\Delta^4} \cdot \sum_{x^5=0}^{x_{max}} p_{x^5} \cdot \sum_{\ddot{\Delta}=0}^{\ddot{\Delta}_{max}} p_{\ddot{\Delta}} \cdot p_{x^6} \cdot p_{next} \tag{5.56}$$
$$\cdot p_{transition} \cdot p_{\Delta^5}$$

The utilization of the server ρ_{server} is determined from the quotient of the sum of the expected values of the time intervals $E(T^2)$, $E(T^4)$ and $E(T^5)$ and the expected value of the cycle time $E(T)$ using the inverse probability:

$$\rho_{server} = 1 - \frac{E(T^2) + E(T^4) + E(T^5)}{E(T)} \tag{5.57}$$

To calculate the utilization of the system, only time gaps in which the system is completely empty are considered. This is the case in time interval Δ^4 and can be possible in time interval Δ^2, but it does not have to be. Accordingly, the part of Δ^2 in which all queues are empty must be derived. If all queues are empty at the beginning of the time gap, the part of the time interval Δ^{21} is equal to the minimum residual interarrival time. Otherwise $\Delta^{21} = 0$ applies. The probability $p_{\Delta^{21}}$ thus can be determined with the system state x^2:

$$\begin{aligned} p_{\Delta^{21}} &= P(T^{21} = \Delta^{21} \mid \Lambda^2 = x^2) \\ &= \begin{cases} 1 & \sum_{n=0}^{N} e^2_n > 0, \quad \Delta^{21} = 0 \\ 1 & \sum_{n=0}^{N} e^2_n = 0, \quad \Delta^{21} = \min_{i \in \{1,\ldots,N\}} \{g^2_i\} \\ 0 & \text{otherwise} \end{cases} \end{aligned} \tag{5.58}$$

$E(T^{21})$ can be calculated equivalent to equation (5.54) based on the steady state distribution $\vec{\lambda}$ and the conditional probabilities p_{Δ^1}, p_{x^2} and $p_{\Delta^{21}}$:

$$E(T^{21}) = \sum_{\Delta^{21}=0}^{\Delta^2_{max}} \Delta^{21} \cdot \sum_{x=0}^{x_{max}} \lambda_x \sum_{\Delta^1=0}^{\Delta^1_{max}} p_{\Delta^1} \cdot \sum_{x^2=0}^{x_{max}} p_{x^2} \cdot p_{\Delta^{21}} \tag{5.59}$$

From the inverse probability of the sum of the expected values of the time intervals $E(T^{21})$ and $E(T^4)$ divided by the expected value of the cycle time $E(T)$ the utilization of the system ρ_{system} can be determined:

$$\rho_{system} = 1 - \frac{E(T^{21}) + E(T^4)}{E(T)} \tag{5.60}$$

5.3.6 Rejection Probabilities

The rejection probability of a queue is determined from the average number of rejections in relation to the average number of arrivals in a cycle. To calculate the expected value of the rejections in a cycle, the transition probabilityp_{x^t} depending on a time interval t must be divided into three parts. First the transition is calculated without capacity restrictions. Then the rejections of a queue are determined. Finally, the unlimited queue lengths are transformed into limited queue lengths. The expected value of the arrivals in a cycle is determined based on the number of customers in a system state and the number of customers in a subsequent system state.

The transition probability of queue length $\bar{e}_i^t$ and residual interarrival time g_i^t of a queue i in a time interval Δ^t without capacity restriction is based on equation (5.6). The five cases considered in Section 5.2.3 are thus reduced to the first three cases:

$$P_i(\bar{Q}_i^{t+1} = \bar{e}_i^{t+1}, R_i^{t+1} = g_i^{t+1} \mid \bar{Q}_i^t = \bar{e}_i^t, R_i^t = g_i^t, T^t = \Delta^t)$$
$$= \begin{cases} 1 & \bar{e}_i^{t+1} = \bar{e}_i^t - \Theta, \quad g_i^t > \Delta^t, \\ & g_i^{t+1} = g_i^t - \Delta^t \\ a_{i,\Delta^t - g_i^t + g_i^{t+1}} & \bar{e}_i^{t+1} = \bar{e}_i^t - \Theta + 1, \quad g_i^t \leq \Delta^t \\ \sum\limits_{m=0}^{m_{max}} a_{i,\Delta^t - g_i^t - m}^{\otimes \bar{e}_i^{t+1} - \bar{e}_i^t - \Theta - 2} \cdot a_{i,m+g_i^{t+1}} & \bar{e}_i^{t+1} > \bar{e}_i^t - \Theta + 1, \quad g_i^t \leq \Delta^t \\ 0 & \text{otherwise} \end{cases} \tag{5.61}$$

In the same way as in Section 5.2.3 the transition probability of queue lengths and residual interarrival times without capacity restrictions $p_{\vec{x}^{t+1}}$ of all queues depending on a time interval Δ^t can be determined by multiplying the individual transition probabilities of the queues $1,\ldots,N$ from equation (5.61):

$$\begin{aligned} p_{\vec{x}^{t+1}} =& P(\bar{\Lambda}^{t+1} = \vec{x}^{t+1} \mid \bar{\Lambda}^t = \vec{x}^t, T^t = \Delta^t) \\ =& P(\bar{Q}_1^{t+1} = \vec{e}_1^{t+1},\ldots,\bar{Q}_N^{t+1} = \vec{e}_N^{t+1}, R_1^{t+1} = g_1^{t+1},\ldots,R_N^{t+1} = g_N^{t+1} \mid \\ & \bar{Q}_1^t = \vec{e}_1^t,\ldots,\bar{Q}_N^t = \vec{e}_N^t, R_1^t = g_1^t,\ldots,R_N^t = g_N^t, T^t = \Delta^t) \\ =& \prod_{i=1}^{N} P_i(\bar{Q}_i^{t+1} = \vec{e}_i^{t+1}, R_i^{t+1} = g_i^{t+1} \mid \bar{Q}_i^t = \vec{e}_i^t, R_i^t = g_i^t, T^t = \Delta^t) \end{aligned} \tag{5.62}$$

The probability of the number of rejections $p_{\eta_i^t}$ of a queue i in a time interval depending on the number of customers in a queue i without capacity restriction can be derived from two cases. If the number of customers in the unlimited queue is less than or equal to K_i, the number of rejections η_i^t is equal to 0. If it is greater than K_i, the number of rejections is calculated from the difference between the number of customers in the queue $\vec{e}_i^t$ and the capacity of the queue K_i. Mathematically this can be formulated as follows:

$$p_{\eta_i^t} = P_i(H_i^t = \eta_i^t \mid \bar{Q}_i^t = \vec{e}_i^t) = \begin{cases} 1 & \vec{e}_i^t \leq K_i, \quad \eta_i^t = 0 \\ 1 & \vec{e}_i^t > K_i, \quad \eta_i^t = \vec{e}_i^t - K_i \\ 0 & \text{otherwise} \end{cases} \tag{5.63}$$

The relationship between the number of customers in a queue i with and without capacity restriction is calculated in the next step to transform the unlimited queue lengths. If the number of customers $\vec{e}_i^t$ of the unlimited queue i is less than or equal to the capacity limit K_i, the number of customers with and without capacity restriction is equal. If it is greater, the number of customers e_i^t of the restricted queue is set to K_i. The transformation from $\vec{e}_i^t$ to e_i^t can then be described according to the equation (5.64).

$$P_i(Q_i^t = e_i^t \mid \bar{Q}_i^t = \bar{e}_i^t) = \begin{cases} 1 & \bar{e}_i^t \leq K_i, \quad e_i^t = \bar{e}_i^t \\ 1 & \bar{e}_i^t > K_i, \quad e_i^t = K_i \\ 0 & \text{otherwise} \end{cases} \tag{5.64}$$

The transition probability of queue lengths and residual interarrival times depending on a time interval for all queues is determined by the mathematical product:

$$\begin{aligned} p_{x^{t+1*}} =& P(Q_1^{t+1} = e_1^{t+1}, \ldots, Q_N^{t+1} = e_N^{t+1} \mid \\ & \bar{Q}_1^{t+1} = \bar{e}_1^{t+1}, \ldots, \bar{Q}_N^{t+1} = \bar{e}_N^{t+1}) = \prod_{i=1}^{N} P_i(Q_i^t = e_i^t \mid \bar{Q}_i^t = \bar{e}_i^t) \end{aligned} \tag{5.65}$$

In order to determine the average rejection probability of a queue, the total number of rejections per queue in a cycle must be calculated. Therefore, the individual numbers of rejections $\eta_i^1, \ldots, \eta_i^5$ of a queue i in the time intervals $\Delta^1, \ldots, \Delta^5$ are combined to a total number of rejections η_i in a cycle by summing the individual numbers:

$$\begin{aligned} p_{\eta_i} &= P_i(H_i = \eta_i \mid H_i^1 = \eta_i^1, \ldots, H_i^5 = \eta_i^5) \\ &= \begin{cases} 1 & \eta_i = \eta_i^1 + \eta_i^2 + \eta_i^3 + \eta_i^4 + \eta_i^5 \\ 0 & \text{otherwise} \end{cases} \end{aligned} \tag{5.66}$$

The expected value of the rejections of a queue in a cycle can now be calculated according to the transition probability equation (5.41). The transition of queue lengths and residual interarrival times per time interval is done by the three conditional probabilities $p_{\bar{x}^{t+1}}$, $p_{\eta_i^t}$ and $p_{x^{t+1*}}$. Using the steady state distribution $\vec{\lambda}$ the expected value of the rejections $E(H_i)$ of a queue i in a cycle can be determined according to the equation (5.67) on the basis of the equations (5.62), (5.63), (5.65) and (5.66).

$$E(H_i) = \sum_{\eta_i=0}^{\left\lceil \frac{\Delta_{max}}{a_{i,max}} \right\rceil} \eta_i \cdot \sum_{x=0}^{x_{max}} \lambda_x \cdot \sum_{y=x^6=0}^{x_{max}} \sum_{\Delta^1=0}^{\Delta^1_{max}} p_{\Delta^1} \cdot \sum_{\bar{x}^2=0}^{\bar{x}_{max}} p_{\bar{x}^2} \cdot \sum_{\eta_i^1}^{\left\lceil \frac{\Delta_1}{a_{i,max}} \right\rceil} p_{\eta_i^1}$$
$$\cdot \sum_{x^2=0}^{x_{max}} p_{x^{2*}} \cdot \sum_{\Delta^2=0}^{\Delta^2_{max}} p_{\Delta^2} \cdot \sum_{\bar{x}^3=0}^{\bar{x}_{max}} p_{\bar{x}^3} \cdot \sum_{\eta_i^2}^{\left\lceil \frac{\Delta_2}{a_{i,max}} \right\rceil} p_{\eta_i^2} \cdot \sum_{x^3=0}^{x_{max}} p_{x^{3*}} \cdot \sum_{\Delta^3=0}^{\Delta^3_{max}} p_{\Delta^3}$$
$$\cdot \sum_{\bar{x}^4=0}^{\bar{x}_{max}} p_{\bar{x}^4} \cdot \sum_{\eta_i^3}^{\left\lceil \frac{\Delta_3}{a_{i,max}} \right\rceil} p_{\eta_i^3} \cdot \sum_{x^4=0}^{x_{max}} p_{x^{4*}} \cdot \sum_{\Delta^4=0}^{\Delta^4_{max}} p_{\Delta^4} \cdot \sum_{\bar{x}^5=0}^{\bar{x}_{max}} p_{\bar{x}^5} \cdot \sum_{\eta_i^4}^{\left\lceil \frac{\Delta_4}{a_{i,max}} \right\rceil} p_{\eta_i^4} \tag{5.67}$$
$$\cdot \sum_{x^5=0}^{x_{max}} p_{x^{5*}} \cdot \sum_{\ddot{\Delta}=0}^{\ddot{\Delta}_{max}} p_{\ddot{\Delta}} \cdot \sum_{\Delta^5=0}^{\Delta^5_{max}} p_{\Delta^5} \cdot \sum_{\bar{x}^6=0}^{\bar{x}_{max}} p_{\bar{x}^6} \cdot \sum_{\eta_i^5}^{\left\lceil \frac{\Delta_5}{a_{i,max}} \right\rceil} p_{\eta_i^5} \cdot \sum_{x^6=0}^{x_{max}} p_{x^{6*}}$$
$$\cdot p_{next} \cdot p_{transition} \cdot p_{\eta_i}$$

In the next step, the expected value of the arrivals in a queue in a cycle must be determined. For this purpose, the probability of the number of arrivals p_{ι_i} in a queue i depending on the system state x and y is defined:

$$\begin{aligned} p_{\iota_i} &= P_i(L_i = \iota_i \mid \Lambda^\tau = x^\tau, \Lambda^{\tau+1} = x^{\tau+1}) \\ &= P_i(L_i = \iota_i \mid Q_i^\tau = e_i^\tau, Q_i^{\tau+1} = e_i^{\tau+1}) \\ &= \begin{cases} 1 & \iota_i = e_i^{\tau+1} - e_i^\tau - \Theta \\ 0 & \text{otherwise} \end{cases} \end{aligned} \tag{5.68}$$

Since the number of arrivals in a queue does not depend on the subintervals, the expected value of the arrivals $E(L_i)$ in a queue i in a cycle can be calculated using the steady state distribution $\vec{\lambda}$, the transition probability p_{xy}:

$$E(L_i) = \sum_{\iota_i=0}^{\left\lceil \frac{\Delta_{max}}{a_{i,max}} \right\rceil} \iota_i \cdot \sum_{x=0}^{x_{max}} \lambda_x \cdot \sum_{y=0}^{x_{max}} p_{xy} \cdot p_{\iota_i} \tag{5.69}$$

Finally, the rejection probability $p_{i,rejection}$ of a queue i can be determined:

$$p_{i,rejection} = \frac{E(H_i)}{E(L_i)} \tag{5.70}$$

5.4 Chapter Conclusion

In this chapter a new mathematical model called *multi-queue system with multiple departure streams (MQSMDS)* was presented. The queueing model consists of N queues, one server and M sinks and can represent different service rules. The MQSMDS is modelled as a discrete time Markov chain. The following distributions and probabilities were calculated in summary:

- transition probabilities $p_{11}, \ldots, p_{1x_{max}}, \ldots, p_{x_{max}x_{max}}$ from the system states $1,...,x_{max}$ to the system states $1,...,x_{max}$,
- distribution of the system states $\vec{\lambda}$ in the steady state,
- distributions of the number of customers $\vec{q}_1, \ldots, \vec{q}_N$ of the queues $1, \ldots, N$,
- distributions of the waiting time of a customer $\vec{w}_1, \ldots, \vec{w}_N$ of the queues $1, \ldots, N$,
- distributions of the interdeparture time $\vec{d}_1, \ldots, \vec{d}_M$ of the sinks $1, \ldots, M$,
- distributions of the sojourn time $\vec{u_{11}}, \ldots, \vec{u_{1M}}, \ldots, \vec{u_{NM}}$ from the queues $1, \ldots, N$ to the sinks $1, \ldots, M$,
- distribution of the total sojourn time $\vec{u}$,
- utilization of the server ρ_{server},
- utilization of the system ρ_{system} and
- rejection probabilities $p_{1,rejection}, \ldots, p_{N,rejection}$ of the queues $1, \ldots, N$.

6 Simulation of a Multi-Queue System with Multiple Departure Streams under Different Service Rules

Just because something doesn't do what you planned it to do doesn't mean it's useless.
-Thomas A. Edison

Using the analytical model presented in Chapter 5, performance parameters of the MQSMDS can be calculated and thus the system behaviour can be investigated. However, the possible calculations with the analytical model are limited due to the large state space. The size of the state space I can be calculated from the number of queues N, the number of sinks M, the capacities K_i and the maximum interarrival time $a_{i,max}$ as follows:

$$I = N \cdot M \cdot \prod_{i=1}^{N} (K_i + 1) \cdot a_{i,max} \tag{6.1}$$

The transition matrix consists of I^2 probability values. Since the steady state probabilities are determined by solving a linear equation system containing the transition matrix, the transition matrix must be calculated and saved in advance. The state space size and thus the transition matrix size shows an exponential increase with an increasing number of queues. Figure 6.1 shows the development of the transition matrix size and the estimated required memory (estimation with 8 bytes per probability value in the transition matrix) depending on N with the constant values $M = 2$, $K_i = 10$ and $a_{i,max} = 10$. The y-axis of the diagrams has a logarithmic scaling for a better illustration. In the example with $N = 3$ a system with $7,99 \cdot 10^6$ system states results. Accordingly, the transition matrix consists of $6.38 \cdot 10^{13}$ probability values, which corresponds to an approximate memory requirement of 475,170GB.

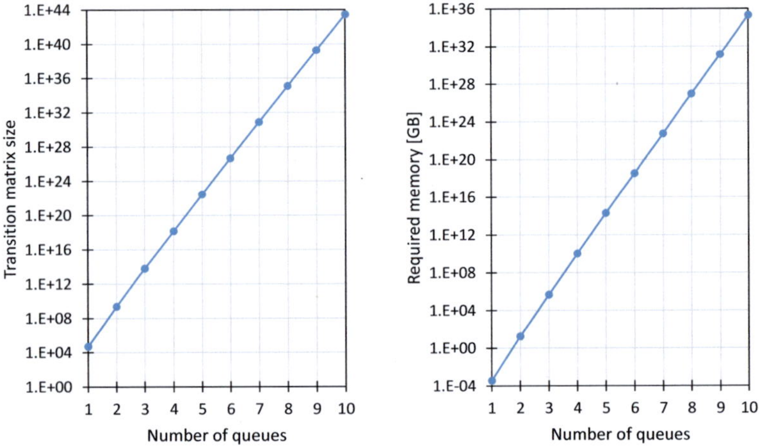

Figure 6.1: Development of the transition matrix size and the required memory for the transition matrix with an increasing number of queues

The analytical model of the MQSMDS was implemented in Java and the experiments were conducted on an Intel(R) Xeon(R) CPU E5-2630 v3 @ 2.40GHz (2 processors) with 64GB RAM. With the available memory, systems with a maximum of 50,000 system states can be calculated. This limits the performance parameter calculations to systems with $N = 2$. For this reason, a simulation model will be developed and validated in the following chapter. Using the simulation model, systems with multiple queues and large distributions (in number of classes) for the interarrival time, processing time and switching time can be investigated.

In Section 6.1, the sequence of the discrete-event simulation of a MQSMDS is illustrated and the individual steps of the simulation model are explained. The simulation model is validated by a comparison with the results of the analytical model. For a set of different parameter configurations the steady state distribution and the performance parameters are determined with the simulation model as well as with the analytical model and the results are compared. The validation of the steady state distribution is based on the chi-squared test and the determination of the relative and absolute deviation on average and in maximum. The performance parameters are validated using the relative and absolute deviation of the expected values and 95%-quantiles. In Section 6.2 the validation approach is explained and the tested parameter configurations are defined. The validation results of the steady state distribution and the performance parameters are presented in Section 6.3. In Section 6.4 the results are summarized.

6.1 Discrete-Event Simulation

In a discrete-event simulation (DES), all events that occur during the execution of a discrete process are simulated. For each event an event routine dependent on the event type is executed (Hedtstück 2013). This discrete-event simulation consists of six events that can or must occur within a simulation iteration. According to Section 5.2.1 a simulation iteration is equivalent to a cycle. The events *end of time gap 1*, *end of switching*, *end of processing*, *end of time gap 2* and *end of time gap 3* must occur in an iteration. Additionally, the event *arrival of a customer* can occur in the iteration. When a customer is generated, his properties queue, sink and interarrival time are assigned and set. A simulation iteration consists of the following event routines of the six events:

1. Arrival of a customer:
   a) Add a customer to the queue that corresponds to the queue of the arriving customer.
   b) Generate the next customer and set his properties:
      i. Set the queue of the customer to the queue where a customer just arrived.
      ii. Determine the arrival time of the customer using a random number based on A.
      iii. Determine the sink of the customer using a random number based on $\hat{P}$.
2. End of switching:
   a) Determine the end of the time gap 1.
3. End of time gap 1:
   a) Determine processing time using a random number based on S.
   b) Remove a customer from the queue to be processed.
4. End of processing:
   a) Determine whether the same queue is selected again.
   b) Determine the end of the time gap 2.
5. End of time gap 2:
   a) Determine whether the same queue is selected again.
   b) Determine iteratively the end of the time gap 3.
6. End of time gap 3:
   a) Determine next queue to be served.
   b) Determine switching time using a random number based on C.

By using a random number generator the random variables arrival time A, switching time C, processing time S and executed transition $\hat{P}$ are approximated. The end of time gap 1 is determined with the arrival time of the first customer of the queue being served. If the arrival time is less than or equal to the current time, the customer has already arrived and the end of the time gap corresponds to the current time. Otherwise, the end of time gap 1 corresponds to the arrival time of the customer.

In order to calculate the end of time gap 2, it is first determined whether the same queue is selected again. The selected rule types of rule category 2 are used to determine whether one of the criteria is infringed. If no additional rule applies for an empty system, the time gap 2 is 0 and accordingly the end of the time gap is identical to the current time (see Section 5.2.1). Otherwise, the end of time gap 2 corresponds to the maximum of the minimum arrival time of all queues and the current time. If a customer is already in a queue, the end of the time gap equates to the current time and otherwise to the earliest arrival time of a customer.

Since time may have elapsed between the end of processing and the end of the time gap 2, it has to be checked whether the same queue is selected once more or not before the end of time gap 3 is determined. For the rule types where the *additional rule for temporally changing decisions* applies (see Section 4.2), the end of time gap 3 is determined iteratively by increasing the time by one time unit. After each time unit it is checked whether the queue is still empty and the criteria to continue serving a queue are fulfilled. If one of the two points no longer applies, the iterative process is terminated and the end of time gap 3 corresponds to the current time. At the end of time gap 3, the next queue to be served is calculated according to the rule type of rule category 1 of the selected service rule. For a stochastically selected queue the next queue to be served is determined with the random generator based on $\tilde{P}$.

The calculation of the warm-up period, the termination criterion of a simulation run and the number of repetitions are based on commonly used rules for discrete-event simulations. To determine the warm-up period the heuristic MSER-5 (marginal standard error rule) is used.

MSER-5 is an algorithm based on batched data (batch size of 5) to find the point in the data series where the standard error in the data is minimal (White 1997). The optimal point in the data series is determined using the number of waiting customers in the queues. The expected values of the total queue length $\tilde{Q}^t$ with $t = 0, \ldots, n$ are determined over a series of 5 time units. Based on the sum of the squared deviations of $\tilde{Q}^t$ from the expected value $E(\tilde{Q}(d,n)) = 1/(n-d) \cdot \sum_{t=d+1}^{n} \tilde{Q}^t$ the optimal truncation point d^* is determined:

$$d^* = \underset{0 \leq d < n}{\arg\min} \left\{ \frac{1}{(n-d)^2} \cdot \sum_{t=d+1}^{n} (\tilde{Q}^t - E(\tilde{Q}(d,n)))^2 \right\} \tag{6.2}$$

According to (Hoad and Robinson 2011) the computation stops calculating the test statistic at a default of 5 batches from the end of the data series. In addition, an iterative calculation of the truncation point is performed, at which the sample is iteratively increased until a valid truncation point is reached. An estimated truncation point is invalid, if it falls into the second half of the data series. In this case, more data is generated (10% more batches of the start sample of the time interval T_1) and analysed until a truncation point is found within the first half of the data series.

The termination of a simulation run is determined based on the relative deviation of the expected value of the steady state distribution from two points of view. If the relative difference at the beginning and end of the analysis interval T_2 is less than ε_1, the simulation run is cancelled.

The number of repetitions m of the simulation run is determined using the confidence interval (Arnold and Furmans 2009). The estimator of the confidence interval for the expected value $E(\Lambda)(m)$ of the steady state distribution calculated from samples is determined assuming an (approximately) normally distributed population based on the t-distribution

$t_{m-1,1-\alpha/2}$ with a probability of error $\alpha = 0.01$ and the empirical variance $S^2(m) = 1/(m-1) \cdot \sum_{i=1}^{m} (\Lambda^i - E(\Lambda(m)))^2$ of the samples $(\Lambda^1, \ldots \Lambda^m)$. If the confidence interval weighted with the unbiased point estimate of $E(\Lambda(m))$ is less than ε_2, the repetition of the simulation run is terminated:

$$\frac{2 \cdot t_{m-1,1-\alpha/2} \cdot \sqrt{\frac{S^2(m)}{m}}}{E(\Lambda(m))} < \varepsilon_2 \tag{6.3}$$

With a small number of samples, it is possible that by chance the certain expected values are close to each other, which leads to a very low confidence interval. This is just an artefact of calculating standard error for a very small sample size. In order to avoid this occurrence a minimum number of repetitions m_{min} is specified.

The simulation can be used to measure observations of the system state and the performance parameters for various parameter configurations. The distributions of these observations can be compared with the results of the analytical model to validate the simulation model.

6.2 Validation Approach

For validation, the results of the simulation model are compared with the results of the analytical model with regard to various parameter configurations. The parameter configurations are divided into the 3 groups *large*, *medium* and *small*. For each group, a minimum (MIN) and maximum (MAX) value is set for the required input parameters. Within these limits, random values for the respective parameters are determined for each configuration. The distributions of the interarrival times, processing times and switching times consist of two randomly determined possible values. The probabilities for these values are also determined randomly. The number of queues is fixed, due to the computing time and the required memory. In the group *large* the different parameters like queue capacity, interarrival, switching and processing times can have higher values. In the parameter configurations of the group *small* the limits are close to each other and the possible values are rather small. In Table 6.1 the limits of the tested parameter configurations for the 3 groups are shown.

Parameters	Notations	Parameter configurations					
		Large		Medium		Small	
		MIN	MAX	MIN	MAX	MIN	MAX
Number of queues	N	2	2	2	2	2	2
Number of sinks	M	1	2	1	2	1	1
Queue capacities	K_1, K_2	2	5	1	3	1	1
Interarrival times	A_1, A_2	4	10	1	5	2	3
Processing times	S_1, S_2	1	3	1	2	1	2
Switching times	$C_1, C2$	0	2	0	1	0	1
Time window	TW_1, TW_2	4	10	2	4	1	3
Number of services in the queues	MN_1, MN_2	2	4	1	3	1	2
Limit values related to queue length	LV_1^1, LV_2^1	2	K_i	1	K_i	1	K_i
Limit values related to waiting time	LV_1^2, LV_2^2	4	10	2	4	1	3

Table 6.1: Minimum and maximum values of the input parameters for the tested parameter configurations of the 3 groups

In order to generate reasonable parameter configurations, it must be ensured that the utilization is less than 1. The utilization of the system can be estimated by the estimated expected values of total arrival time $E(\tilde{A})$, total processing time $E(\tilde{S})$ and total switching time $E(\tilde{C})$ independent of the service rule. The calculation is based on the assumptions that the server is only empty if there is no customer in the system, the queues are without capacity restrictions and the expected value of the waiting time and the service time are independent of each other (see Kleinrock's conservation law, Section 3.2.1). Since these assumptions are not valid in the considered model, the calculation of the estimated utilization of the system $\tilde{\rho}$ is only an approximation. The expected values of arrival, switching and service times for all queues and sinks are estimated by weighting the expected values of the individual queues and sinks with the throughput of the individual queues and the transition probability. Also in this calculation, rejections and influences on the expected values due to special service rules are neglected, so that the calculation of $E(\tilde{A})$, $E(\tilde{S})$ and $E(\tilde{C})$ is approximative:

$$\tilde{\rho} = \frac{E(\tilde{S}) + E(\tilde{C})}{E(\tilde{A})} \tag{6.4}$$

with

$$E(\tilde{A}) = \frac{1}{\sum\limits_{i=1}^{N} \frac{1}{E(A_i)}} \tag{6.5}$$

$$E(\tilde{S}) = \sum_{i=1}^{N} \sum_{j=1}^{M} \frac{E(\tilde{A})}{E(A_i)} \cdot \left(\hat{p}_{i,j} \cdot E(S_{ij})\right) \tag{6.6}$$

$$E(\tilde{C}) = \sum_{j=1}^{M} \sum_{i=1}^{N} \frac{\sum\limits_{n=1}^{N} \hat{p}_{n,j}}{N} \cdot \left(\frac{E(\tilde{A})}{E(A_i)} \cdot E(C_{ji}) \right) \tag{6.7}$$

For each group, a low ($0.5 \leq \tilde{\rho} < 0.7$), medium ($0.7 \leq \tilde{\rho} < 0.9$) and high ($0.9 \leq \tilde{\rho} < 1.0$) utilization level is considered, based on the estimated utilization. After the parameters have been set for a configuration using random values, the system checks whether the utilization corresponds to the required utilization level. If this is not the case, the parameters are reset and new random values are generated. This is repeated until the required utilization level is reached. For each group and utilization level 10 parameter configurations are created. The generated parameter configurations can be found in appendix C.1 in Table C.1.

The calculation of the resulting parameters for all 480 service rules and for each configuration results in a total of 43,200 possible test scenarios. However, with the available memory, the possible state space to be calculated is limited to 50,000 (see Figure 6.1). If this limit of the state space for a configuration is exceeded for a service rule, the calculation is cancelled for this service rule and is continued with the next. For the validation, the steady state distribution and all performance parameters are determined and saved for each configuration and service rule.

The required parameters for determining the warm-up period, the termination of a simulation run and the number of repetitions are determined by previous simulation experiments and set to $T_1 = 1000$, $T_2 = 1000$, $\varepsilon_1 = 1 \cdot 10^{-7}$, $\varepsilon_2 = 0.0025$ and $n_{min} = 10$.

6.3 Validation Results

The steady state distribution, the distributions of the number of customers of the queues, the utilizations for server and system and the rejection probabilities could be calculated for 22,429 of the 43,200 possible test scenarios. When calculating the distributions of the waiting time of a customer, the distributions of the interdeparture time and the distributions of the sojourn time, the state space is significantly larger due to the extended state spaces and they can only be determined for configurations of the group *small*. Here, 12,176 calculations of the waiting time and sojourn time distributions and 10,244 calculations of the interdeparture time distributions could be performed out of 14,400 possible scenarios.

The average warm-up time over the simulation runs of the parameter configurations is 111.28 time units and the average number of observations in a simulation run is 423,136.31. On average, 18.17 repetitions of a simulation run were performed.

6.3.1 Validation of the Steady State Distribution

The assumption of having achieve a steady state distribution is validated by a chi-square test and the calculation of the relative and absolute deviation in the mean and maximum. The results are summarized in the Table 6.2.

Based on a chi-square test, the hypothesis H_0 that the system states that occurred in the simulation model are independent and identically distributed random variables of the steady state distribution of the analytical model is considered. The hypothesis is tested for the significance level $\alpha = 0.01$ and $\alpha = 0.05$. The determination of the number of classes K and thus the minimum probability mass of a class is based on the conditions for an appropriate classification with $K \leq 5 \cdot \lg e_i$, $K \geq 3$ and $e_i > 1$, $\forall i \in \{1, \ldots, K\}$ where e_i corresponds to the number of observations in class i (Law et al. 2015). The class division is based on the first fit decreasing algorithm, which is used to solve bin packing problems (Korte and Vygen 2012). Since the indexing of the steady states results in an arbitrary number, the order of the indices does not matter and can be changed by redistributing the probabilities to the classes. Based on the first fit decreasing algorithm, a uniform distribution of the probabilities over the classes can be achieved, which is also required for the chi-square test (Law et al. 2015).

The results in Table 6.2 show that the null hypothesis for all tested parameter configurations for the significance level $\alpha = 0.01$ for 91.95% and for the significance level $\alpha = 0.05$ for 89.11% of the tested parameter configurations is not rejected and therefore confirms the validity of the simulation model.

For the parameter configurations *large* (see Table 6.1) with a low utilization ($0.5 \leq \tilde{\rho} < 0.7$) 81.19% are not rejected for the significance level $\alpha = 0.01$. The higher rejection rate can be explained by rarely occurring states that do not occur in the simulation.

Validation tests	Validation parameters	Parameter configurations								
		Large			Medium			Small		
		$0.5 \leq \tilde{\rho} < 0.7$	$0.7 \leq \tilde{\rho} < 0.9$	$0.9 \leq \tilde{\rho} < 1.0$	$0.5 \leq \tilde{\rho} < 0.7$	$0.7 \leq \tilde{\rho} < 0.9$	$0.9 \leq \tilde{\rho} < 1.0$	$0.5 \leq \tilde{\rho} < 0.7$	$0.7 \leq \tilde{\rho} < 0.9$	$0.9 \leq \tilde{\rho} < 1.0$
Chi-squared test	Number of classes	24.17	22.33	22.25	24.06	22.41	13.91	8.81	7.04	5.07
	Minimum class probability mass	0.12	0.07	0.07	0.07	0.06	0.10	0.18	0.18	0.24
	Pearson's cumulative test statistic	37.30	24.34	23.34	26.90	22.93	14.46	9.48	8.62	14.46
	χ^2 value for 0.99 of the CDF	40.79	38.88	38.77	41.23	39.14	27.00	19.55	18.42	10.83
	χ^2 value for 0.95 of the CDF	34.60	32.72	32.62	34.90	32.93	21.84	15.08	14.08	7.39
	H_0 is not rejected for $\alpha = 0.01$	81.19%	97.67%	98.61%	97.21%	99.85%	98.95%	97.38%	98.71%	70.09%
	H_0 is not rejected for $\alpha = 0.05$	77.51%	97.01%	98.46%	95.21%	99.41%	98.25%	95.61%	97.25%	61.26%
	p-value	0.26	0.42	0.51	0.32	0.42	0.40	0.38	0.36	0.24
Deviation test	Average relative deviation	$3.07 \cdot 10^{-3}$	$2.89 \cdot 10^{-3}$	$1.91 \cdot 10^{-3}$	$3.77 \cdot 10^{-3}$	$4.01 \cdot 10^{-3}$	$2.34 \cdot 10^{-3}$	$1.24 \cdot 10^{-3}$	$1.16 \cdot 10^{-3}$	$1.31 \cdot 10^{-3}$
	Maximum relative deviation	$2.69 \cdot 10^{-1}$	$4.03 \cdot 10^{-1}$	$9.99 \cdot 10^{-1}$	$2.44 \cdot 10^{-1}$	$3.18 \cdot 10^{-1}$	$2.71 \cdot 10^{-1}$	$1.67 \cdot 10^{-1}$	$1.64 \cdot 10^{-1}$	$1.11 \cdot 10^{-1}$
	Average absolute deviation	$8.77 \cdot 10^{-7}$	$1.17 \cdot 10^{-6}$	$1.24 \cdot 10^{-6}$	$3.18 \cdot 10^{-6}$	$2.48 \cdot 10^{-6}$	$2.98 \cdot 10^{-6}$	$1.08 \cdot 10^{-5}$	$9.25 \cdot 10^{-6}$	$1.34 \cdot 10^{-5}$
	Maximum absolute deviation	$1.00 \cdot 10^{-4}$	$2.54 \cdot 10^{-4}$	$1.64 \cdot 10^{-4}$	$1.80 \cdot 10^{-4}$	$2.12 \cdot 10^{-4}$	$3.01 \cdot 10^{-4}$	$3.67 \cdot 10^{-4}$	$3.62 \cdot 10^{-4}$	$6.84 \cdot 10^{-4}$

Table 6.2: Results of the validation of the steady state distribution

In parameter configuration 1 with service rule 2, for example, there are 2.352 states with an analytically calculated steady state probability greater than 0 that did not occur in the simulation. The mean steady state probability of these states is $2.18 \cdot^{-9}$. Due to the large number of these states the sum of the probabilities is $5.13 \cdot 10^{-6}$ which leads with 5,596,570 observed system states in the simulation to an increase of the Pearson's cumulative test statistic of 28.72. Together with the small deviations of the simulation results due to simulation variations, this leads to a higher number of rejections.

With a high utilization ($0.9 \leq \tilde{\rho} < 1.0$) and a parameter configuration *small*, a higher percentage of rejections also occurs (29.91% for $\alpha = 0.01$). This can be explained by the fact that in a system with a high utilization only a few system states with high probabilities will occur. For example, in parameter configuration 90 with service rule 2 there are 30 possible states and the cumulated probability of the most frequent six steady states is 0.89. A uniform class division is not possible and the number of classes must be small to fulfil the conditions of an appropriate classification. Accordingly, a minimal deviation of the analytical and simulatively determined probabilities in one of the classes leads to high absolute differences due to the multiplication of the high probability mass of the class with the number of observations and thus to a significantly higher Pearson's cumulative test statistic.

However, if the rejected cases are calculated repeatedly with the simulation model and another chi-square test is performed, the null hypothesis is not rejected again in 67.40% of these cases. Thus it can be assumed that the rejections are a type I error (rejection of a true null hypothesis). Overall, the results of the chi-square test, with the exception of a few non-reproducible individual cases, do not speak significantly against the null hypothesis H_0 and thus not significantly against the assumption that the occurring system of the simulation model originate from the steady state distribution $\vec{\lambda}$ of the analytical model of the MQSMDS. The chi-square test confirms the validity of the developed simulation model in respect to the steady state distribution.

The relative and absolute deviations on average and the maximum values are marginally small. The mean relative deviation amounts to $2.18 \cdot 10^{-3}$ and the mean absolute deviation is $7.63 \cdot 10^{-6}$. The maximum relative deviation averaged over all tested parameter configurations is $9.78 \cdot 10^{-1}$ and the maximum absolute deviation amounts to $5.28 \cdot 10^{-4}$. The somewhat higher relative deviation is to be explained by the partially occurring very small probabilities of some system states. Small absolute deviations lead to higher deviations in a relative sense. Overall, the deviations between the analytical and simulation model are insignificantly small, so that the comparison confirms the validity of the simulation model related to the steady state.

6.3.2 Validation of the Performance Parameters

The validation of the performance parameters is based on the relative deviation ($|\Delta^{rel}|$) and absolute deviation ($|\Delta^{abs}|$) of the expected value and the 95%-quantile. The distributions of the number of customers and the distribution of the waiting time of a customer are considered separately for each queue. The distribution of the interdeparture time is evaluated for each sink and the sojourn time distribution for each source to sink connection and in total. For the utilization, the deviation of the utilization of the server and of the system is determined. The rejection probability deviation is considered per queue. In Table 6.3 the results of the validation of the performance parameters are summarized.

For all performance parameters, the relative and absolute deviations of the expected values are marginally small. The deviation of the 95%-quantile of the performance parameter distributions is 0 in 99.91% of the cases, due to the integer values of the quantiles. The expected value of the performance parameter distribution deviates on average by less than 0.0014 relative and 0.0003 absolute.

Performance parameter	Performance measure	$\lvert\Delta^{rel}\rvert$	$\lvert\Delta^{abs}\rvert$
Number of customers of a queue	$E(\tilde{Q}_1)$	$3.60 \cdot 10^{-4}$	$1.99 \cdot 10^{-4}$
	$\tilde{Q}_{1,0.95}$	$5.94 \cdot 10^{-4}$	$8.92 \cdot 10^{-5}$
	$E(\tilde{Q}_2)$	$3.94 \cdot 10^{-4}$	$2.15 \cdot 10^{-4}$
	$\tilde{Q}_{2,0.95}$	$0.00 \cdot 10^{-0}$	$0.00 \cdot 10^{-0}$
Waiting time of a customer of a queue	$E(W_1)$	$1.66 \cdot 10^{-3}$	$5.92 \cdot 10^{-4}$
	$W_{1,0.95}$	$2.35 \cdot 10^{-4}$	$1.23 \cdot 10^{-3}$
	$E(W_2)$	$1.71 \cdot 10^{-3}$	$6.03 \cdot 10^{-4}$
	$W_{2,0.95}$	$8.35 \cdot 10^{-4}$	$2.05 \cdot 10^{-3}$
Interdeparture time of a sink	$E(D_1)$	$6.43 \cdot 10^{-5}$	$1.04 \cdot 10^{-4}$
	$D_{1,0.95}$	$1.90 \cdot 10^{-3}$	$4.88 \cdot 10^{-3}$
Sojourn time from a queue to a sink	$E(U_{11})$	$3.76 \cdot 10^{-4}$	$6.40 \cdot 10^{-4}$
	$U_{11,0.95}$	$6.02 \cdot 10^{-5}$	$2.46 \cdot 10^{-4}$
	$E(U_{21})$	$3.93 \cdot 10^{-4}$	$6.37 \cdot 10^{-4}$
	$U_{21,0.95}$	$5.20 \cdot 10^{-4}$	$1.23 \cdot 10^{-3}$
Total sojourn time	$E(U)$	$3.76 \cdot 10^{-4}$	$6.40 \cdot 10^{-4}$
	$U_{0.95}$	$6.02 \cdot 10^{-5}$	$2.46 \cdot 10^{-4}$
Utilization probability	ρ_{server}	$1.16 \cdot 10^{-4}$	$8.84 \cdot 10^{-5}$
	ρ_{system}	$1.01 \cdot 10^{-4}$	$8.17 \cdot 10^{-5}$
Rejection probabilities of a queue	$p_{1,rejection}$	$4.44 \cdot 10^{-3}$	$5.58 \cdot 10^{-5}$
	$p_{2,rejection}$	$4.61 \cdot 10^{-3}$	$5.96 \cdot 10^{-5}$

Table 6.3: Results of the validation of the performance parameters

Figure 6.2 shows the distributions of the performance parameters for one example (parameter configuration 90, service rule 2). The marginally small deviations of the expected values on average of less than 0.0032 relative and 0.0004 absolute are not graphically visible. Overall, the deviations with respect to the performance parameters are negligibly small, so that the comparison of the analytical model and the simulation model of a MQSMDS with respect to the performance parameters confirms the validity of the simulation model.

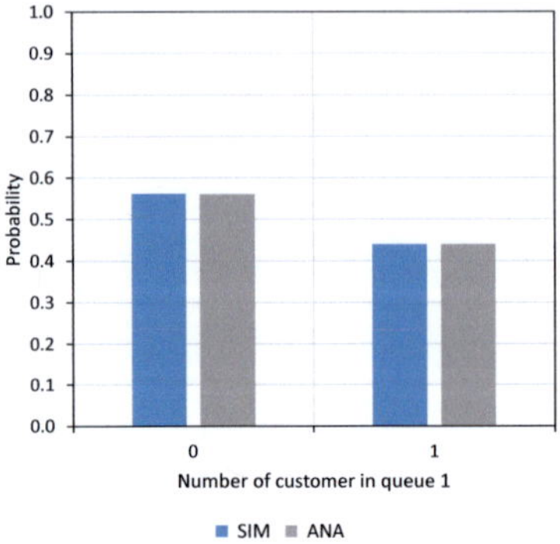
Probability
Number of customer in queue 1
SIM ANA

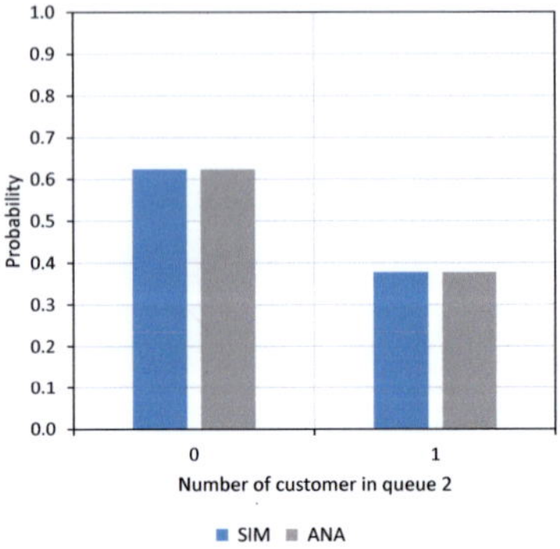
Probability
Number of customer in queue 2
SIM ANA

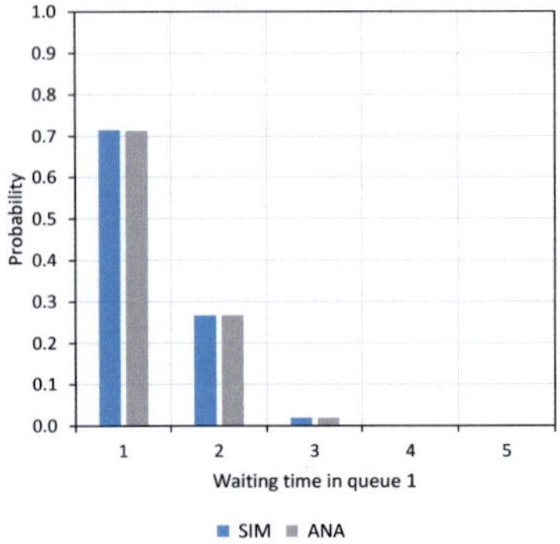
Probability
Waiting time in queue 1
SIM ANA

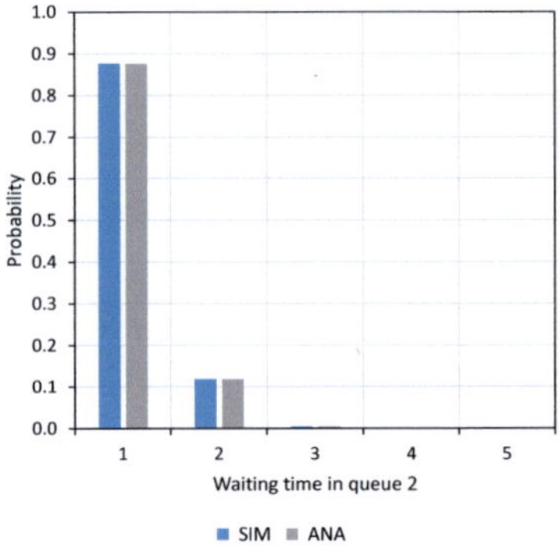
Probability
Waiting time in queue 2
SIM ANA

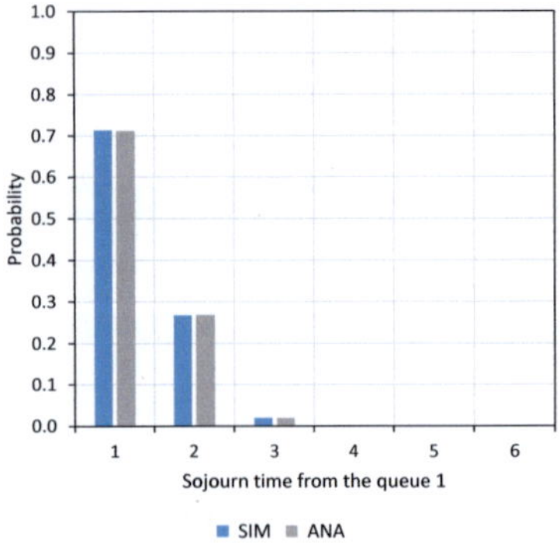
Probability
Sojourn time from the queue 1
SIM ANA

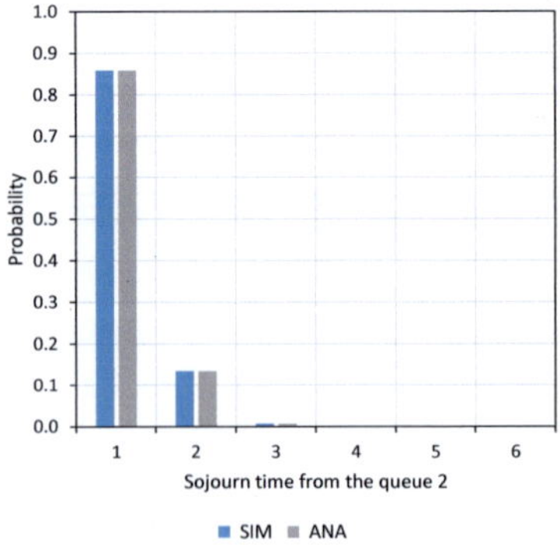
Probability
Sojourn time from the queue 2
SIM ANA

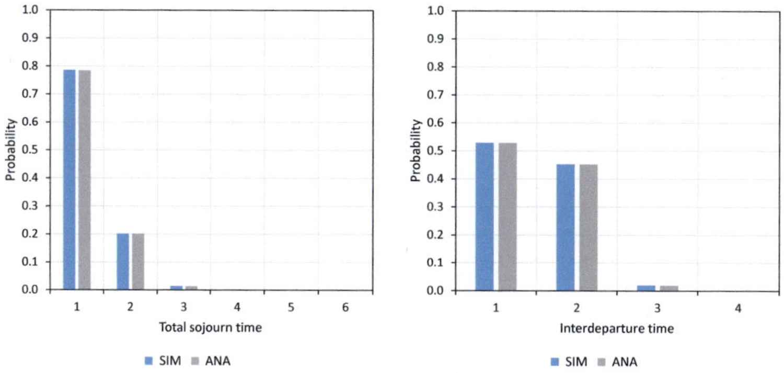

Figure 6.2: Distributions of the performance parameters for parameter configuration 90 and service rule 2

6.4 Chapter Conclusion

In this chapter a simulation model is developed. It represents the MQSMDS using six events that can or must occur within a simulation iteration. This simulation model is necessary due to the limitation of the calculations with the analytical model caused by the large state space. The validity of the simulation model of the MQSMDS is examined on the basis of a comparison with the analytical model. In the chi-square test of the steady state distribution the null hypothesis for all tested parameter configurations for the significance level $a = 0.01$ for 91.95% and for the significance level $a = 0.05$ for 89.11% is not rejected. The relative and absolute deviations between the simulation model and the analytical model with respect to the steady state distribution and the performance parameters are negligibly small. Overall, the comparison of the simulation model and the analytical model of a MQSMDS confirms the validity of the simulation model. Thus, the developed simulation model can be used in a numerical evaluation to analyse the system characteristics and the service rules.

7 Numerical Evaluation

Number rules the universe.

-Pythagoras

In the following chapter, the system characteristics of the MQSMDS are analysed using a numerical study and the service rules are evaluated. The numerical evaluation is carried out using the discrete-event simulation as described in Section 6.1. This allows to consider systems with multiple queues and large distributions (in number of classes) for the interarrival time, processing time and switching time. By varying single or several dependent input parameters, the system behaviour of the MQSMDS is analysed and general statements are made on how the performance parameters behave in relation to these parameters. By comparing the different service rules, these can be evaluated and recommendations can be given regarding their usability.

In Section 7.1 the parameter settings of the simulation experiment are explained. The results of the numerical study are presented in Section 7.2. The numerical study is divided into the analyses of the system characteristics (Section 7.2.1), the evaluation of the service rules (Section 7.2.2) and the recommendations of service rules depending on parameters and objectives (Section 7.2.3). The conclusion of the chapter is presented in Section 7.3.

7.1 Evaluation Approach

The numerical study is performed using discrete-event simulation. The required parameters for determining the warm-up period, termination of a simulation run and number of repetitions are determined by previous simulation experiments and set to $T_1 = 100{,}000$, $T_2 = 1{,}000$, $\varepsilon_1 = 5 \cdot 10^{-4}$, $\varepsilon_2 = 1 \cdot 10^{-3}$ and $m_{min} = 10$. The termination criterion of the simulation run and of the number of repetitions is the expected value and the variability of the interarrival time instead of the expected value of the steady state distribution.

To reflect a broad variety of different settings, the random variables whose distribution is specified as input in the MQSMDS (see Section 5.1) are described systemically using several parameters. The considered distributions of the random variables are in part based on continuous probability distributions, which are discretized by class formation of class width 1. Since the random variables may vary for each queue, sink, or both, the differences of the distributions across the queues or sinks are described by straight lines with given gradients. In addition, special characteristics such as the transition from a queue to a sink with the same number are represented by an additional boolean parameter. Table 7.1 shows the input parameters and their specification for each type of random variable as well as the necessary system parameters and the additional parameters that are required depending on the service rule.

The distribution of transition probabilities D_P across the sinks may be uniform (1), deterministic (2) or unequal (3). With a uniform distribution, the transition probability $p_{i,j}^{(1)}$ from a queue $i \in \{1,\ldots,N\}$ to a sink $j \in \{1,\ldots,M\}$ can be calculated independently of i and j as follows:

$$p_{i,j}^{(1)} = \frac{1}{M} \tag{7.1}$$

Category	Parameters	Notation	Specification
System	Number of queues	N	natural number
	Number of sinks	M	natural number
	Queue capacities per queue	K_i	natural numbers
Transition	Distribution of the probability across the sinks	$D_{\hat{P}}$	uniform, unequal, deterministic
	Gradient of the probability across the sinks	$b_{\hat{P}}$	real numbers
	Highest probability for the transition from a queue to the sink with the same number	BV_1	boolean number
Inter-arrival times	Type of distribution	T_A	uniform, triangular, gamma, discrete
	Expected value	$E(\tilde{A})$	natural number
	Variability of the distribution per queue	$c^2(A_i)$	natural number
	Distribution of the expected value across the queues	D_A	uniform , unequal
	Gradient of the expected value across the queues	b_A	real number
Processing times	Type of distribution	T_S	uniform, triangular, gamma, discrete
	Expected value	$E(\tilde{S})$	natural number
	Variability of the distribution per queue and sink	$c^2(S_{ij})$	natural number
	Distribution of the expected value across the queues	D_{S_i}	uniform, unequal
	Gradient of the expected value across the queues	b_{S_i}	real number
	Distribution of the expected value across the sinks	D_{S_j}	uniform, unequal
	Gradient of the expected value across the sinks	b_{S_j}	real number
Switching times	Type of distribution	T_C	uniform, triangular, gamma, discrete
	Expected value	$E(\tilde{C})$	natural number
	Variability of the distribution per sink and queue	$c^2(C_{ji})$	natural number
	Distribution of the expected value across the queues	D_{C_i}	uniform, unequal
	Gradient of the expected value across the queues	b_{C_i}	real number
	Distribution of the expected value across the sinks	D_{C_j}	uniform, unequal
	Gradient of the expected value across the sinks	b_{C_j}	real number
	No switching time when switching from a sink to the queue with the same number	BV_2	boolean number
Service rule	Type of priority distribution	D_{PN}	ascending, descending
	Type of random distribution	$D_{\tilde{P}}$	uniform, unequal
	Time window per queue	TW_i	natural number
	Maximum number of services per queue	MN_i	natural number
	Limit value of the queue length per queue	LV_i^1	natural number
	Limit value of the waiting time per queue	LV_i^2	natural number
	Service rule number	SR	1-480

Table 7.1: Input parameters for the numerical evaluation of the MQSMDS

Is the transition probability deterministic ($p^{(2)}_{i,j}$), the following applies:

$$p^{(2)}_{i,j} = \begin{cases} 1 & i = j \\ 0 & \text{otherwise} \end{cases} \tag{7.2}$$

If the transition distribution is unequal, the inequality is defined by the gradient of the transition probability across the sinks. In addition, a boolean variable $BV_1 \in \{0,1\}$ determines whether the highest transition probability should be in the diagonal of the transition matrix. The transition probability $p^{(3)}_{i,j}$ from a queue $i \in \{1,\ldots,N\}$ to a sink $j \in \{1,\ldots,M\}$ can then be calculated from an equation describing a straight line equation with the gradient b_P and the constant y_P:

$$p^{(3)}_{i,j} = \begin{cases} (j-1)\cdot b_P + y_P & BV_1 = 0 \\ (i-j-1)\cdot b_P + y_P & BV_1 = 1, \quad i > j \\ (M+i-j-1)\cdot b_P + y_P & BV_1 = 1, \quad i \leq j \\ 0 & \text{otherwise} \end{cases} \tag{7.3}$$

Since the sum of the transition probabilities over all sinks is equal to 1 ($\sum_{j=1}^{M}(j-1)\cdot b_P + y_P = 1$), the constant of the straight line equation can be calculated:

$$y_P = \frac{2 - b_P \cdot M \cdot (M-1)}{2 \cdot M} \tag{7.4}$$

The interarrival time can be distributed uniformly, triangularly, discretely or gamma-distributed. In addition to the expected value $E(\tilde{A})$ and the variability $c^2(\tilde{A})$ of the total interarrival time, the distribution of the expected value over the queues D_A is set as a parameter. A distinction is made between a uniform (1) and an unequal (2) distribution. With a uniform distribution, the expected values of the interarrival time $E(A^{(1)}_i)$ with $i \in \{1,\ldots,N\}$ are equal over the queues and can be determind according to equation (7.5).

$$E(A_i^{(1)}) = E(\tilde{A}) \cdot N \tag{7.5}$$

In the case of an unequal distribution, a straight line equation with a gradient b_A and constant y_A is used. The constant y_A can be determined by inserting the straight line function $(i-1) \cdot b_A + y_A$ into equation (6.5) and transforming this equation. The expected value of the interarrival time $E(A_i^{(2)})$ of a queue $i \in \{1, \ldots, N\}$ can be determined from the straight line equation:

$$E(A_i^{(2)}) = (i-1) \cdot b_A + y_A \tag{7.6}$$

As with the interarrival time, the distribution type of the processing time distinguishes between uniform, triangular, gamma and discrete. The expected value $E(\tilde{S})$, the variability $c^2(S)$ and the distribution of the expected value over the queues D_{S_i} with the gradient b_{S_i} are given in the same way as input parameters. Since the expected value of the processing time can also differ for each sink, a distribution of the expected value over the sinks D_{S_j} is also considered. Again, the unequal distribution over the sinks is done using a straight line function with a given gradient b_{S_j}. The distributions of the expected value over the queues or sinks can be both of type uniform ((1),(1)), unequal over the queues and uniform over the sinks ((2),(1)), uniform over the queues and unequal over the sinks ((1),(2)) or both unequal ((2),(2)). If the expected value is distributed uniformly over the queues and the sinks, the expected value of the processing time $E(S_{ij}^{(1),(1)})$ with $i \in \{1, \ldots, N\}$ and $j \in \{1, \ldots, M\}$ corresponds to the expected value of the total processing time $E(\tilde{S})$ for all queues and all sinks:

$$E(S_{ij}^{(1),(1)}) = E(\tilde{S}) \tag{7.7}$$

In case of an unequal distribution of the processing time over the queues and a uniform distribution over the sinks, the constant y_{S_i} is determined based on the equation (6.6) by inserting the straight line function $(i-1) \cdot b_{S_i} + y_{S_i}$

for $E(S_{ij})$ and transforming the equation to y_{S_i}. The expected value of the processing time $E(S_{ij}^{(2),(1)})$ of a queue $i \in \{1,\ldots,N\}$ can then be calculated independently of the sink $j \in \{1,\ldots,M\}$ as follows:

$$E(S_{ij}^{(2),(1)}) = (i-1) \cdot b_{S_i} + y_{S_i} \tag{7.8}$$

In the same way, the expected value of the processing time $E(S_{ij})$ can be calculated for an unequal distribution over the sinks and a uniform distribution over the queues. The constant y_{S_j} is determined with the equation (6.6). Independently of the queue $i \in \{1,\ldots,N\}$ the expected value of the processing time $E(S_{ij}^{(1),(2)})$ to a sink $jeight \in \{1,\ldots,M\}$ can be calculated:

$$E(S_{ij}^{(1),(2)}) = (j-1) \cdot b_{S_j} + y_{S_j} \tag{7.9}$$

If the expected value is distrThe condition that this additional parameter can be applied is that $N = M$ applies.ibuted unequally over both the queues and the sinks, the expected value of the processing time $E(S_{ij}^{(2),(2)})$ is determined by averaging $E(S_{ij}^{(2),(1)})$ and $E(S_i j^{(1),(2)})$:

$$E(S_{ij}^{(2),(2)}) = \frac{E(S_{ij}^{(2),(1)}) + E(S_{ij}^{(1),(2)})}{2} \tag{7.10}$$

The parameters of the switching time are equivalent to those of the processing time. The constants y_{C_i} and y_{C_j} are determined by the equation (6.7). Additionally, there is a boolean variable BV_2 which specifies whether there is a switching time between queues and sources of the same number or not. The condition that this additional parameter can be applied is that $N = M$. A total of eight different cases result for the calculation of the expected value of the switching time from a sink j to a queue i. If $BV_2 = 1$ applies, the expected value of the switching time from a sink to a queue is calculated using an equation with the case distinction $i = j$ and $i \neq j$. The calculations can be found in the appendix B.5 in the equations (B.25)-(B.28).

For the different distributions of the performance parameters as well as for the distributions of the input parameters the expected value, the variability and the 95%-quantile are measured as output parameters by the simulation. The interarrival time, processing time and switching time are also considered, because they can differ from the input distributions due to rejections and specific service rules. Table 7.2 summarizes the output parameters measured for the numerical evaluation.

Category	Parameter	Notation per queue $\forall i \in \{1,\dots,N\}$	Notation total
Interarrival time	Expected value	$E(A_i)$	$E(A)$
	Variability	$c^2(A_i)$	$c^2(A)$
	95%-quantile	$A_{i,0.95}$	$A_{0.95}$
Processing time	Expected value	$E(S_i)$	$E(S)$
	Variability	$c^2(S_i)$	$c^2(S)$
	95%-quantile	$S_{i,0.95}$	$S_{0.95}$
Switching time	Expected value	$E(C_i)$	$E(C)$
	Variability	$c^2(C_i)$	$c^2(C)$
	95%-quantile	$C_{i,0.95}$	$C_{0.95}$
Number of waiting customers	Expected value	$E(\tilde{Q}_i)$	$E(\tilde{Q})$
	Variability	$c^2(\tilde{Q}_i)$	$c^2(\tilde{Q})$
	95%-quantile	$\tilde{Q}_{i,0.95}$	$\tilde{Q}_{0.95}$
Waiting time of a customer	Expected value	$E(W_i)$	$E(W)$
	Variability	$c^2(W_i)$	$c^2(W)$
	95%-quantile	$W_{i,0.95}$	$W_{0.95}$
Sojourn time	Expected value	$E(U_i)$	$E(U)$
	Variability	$c^2(U_i)$	$c^2(U)$
	95%-quantile	$U_{i,0.95}$	$U_{0.95}$
Utilization probability	Probability of the server	-	ρ_{server}
	Probability of the system	-	ρ_{system}
Rejection probability	Probability	$p_{i,rejection}$	$p_{rejection}$

Table 7.2: Output parameters for the numerical evaluation of the MQSMDS

Category	Objective	Target function
Performance	Maximal throughput of all customers	$\max\left(\frac{1}{E(U)}\right)$
	Maximal throughput for the customers of a specific queue k	$\max\left(\frac{1}{E(U_k)}\right)$
Time Saving	Minimal mean sojourn times of all customers	$\min(E(U))$
	Minimal 95%-quantile of the sojourn times of all customers	$\min(U_{0.95})$
	Minimal mean sojourn times for the customers of a specific queue k	$\min(E(U_k))$
	Minimal 95%-quantile of the sojourn times for the customers of a specific queue k	$\min\left(U_{k,0.95}\right))$
	Minimal mean waiting times of all customers	$\min(E(W))$
	Minimal 95%-quantile of the waiting times of all customers	$\min(W_{0.95})$
	Minimal mean waiting times for the customers of a specific queue k	$\min(E(W_k))$
	Minimal 95%-quantile of the waiting times for the customers of a specific queue k	$\min\left(W_{k,0.95}\right))$
Reduction of queueing	Minimal mean number of waiting customers of all queues	$\min(E(Q))$
	Minimal 95%-quantile of the number of waiting customers of all queues	$\min(Q_{0.95})$
	Minimal maximal mean number of waiting customers over all queues	$\min\left(\max\left(E(\tilde{Q}_i)\right)\right)$
	Minimal maximal 95%-quantile of the number of waiting customers over all queues	$\min\left(\max\left(\tilde{Q}_{i,0.95}\right)\right)$
	Minimal mean number of waiting customers of a specific queue k	$\min\left(E(\tilde{Q}_k)\right)$
	Minimal 95%-quantile of the number of waiting customers of a specific queue k	$\min\left(\tilde{Q}_{k,0.95}\right))$
	Minimal percentage number of rejections for all queues	$\min\left(p_{rejection}\right)$
	Minimal percentage number of rejections for specific queue k	$\min\left(p_{k,rejection}\right)$

Table 7.3: Possible objectives for selecting a service rule

A service rule can be used to achieve different objectives. Based on (Gudehus and Kotzab 2012) Table 7.3 shows the possible objectives with the respective target function. The three categories Performance, Time Saving and Reduction of queueing can be distinguished. In the respective categories, the expected value or the 95%-quantile of the performance parameters is optimized. A total of 18 different target functions can be identified for which the appropriate service rules must be found.

In the following section the system characteristics of the MQSMDS are examined by variation of single and combined input parameters and consideration of the output parameters. The analyses are based on a MQSMDS with five queues and sinks. The parameters of the basic setting can be found in the appendix C.2 in Table C.2. The input parameters are varied in combination whenever a relationship can be assumed between the parameters. Based on the analyses of the system characteristics and additional service rule specific analyses an evaluation of the service rules is derived. The final evaluation is based on the identified influencing parameters and the possible objectives in Table 7.3.

7.2 Evaluation Results

Altogether 630 different parameter settings are simulated and evaluated. The variation of the parameters is based on the basic setting in Table C.2 in the appendix C.2.

The average warm-up time over the simulation runs of the parameter configurations is 538.45 time units and the average number of observations of a simulation run is 9,062,579. On average, 10.76 repetitions of a simulation run were performed.

7.2.1 Evaluation of the System Characteristics

The evaluation of the system characteristics can be separated into the system behaviour in relation to the interarrival time, processing time and switching time. The transition probabilities are not considered further, as they are only an amplifying factor in dependence on processing and switching time. Different distribution types and distributions across the queues and sinks as well as different numbers of queues, utilizations and capacities of the queues are analyzed with respect to various variabilities. The additional parameters required depending on the service rule are considered during the evaluation of the service rules.

Figures 7.1-7.8 shows the results of the evaluation of the system characteristics in relation to the interarrival time. Since the expected value of the total number of waiting customers in the system, the waiting time and the sojourn time of a customer in the system behave similarly with the examined parameters setting, only the average total queue length is shown as a performance parameter in the diagrams.

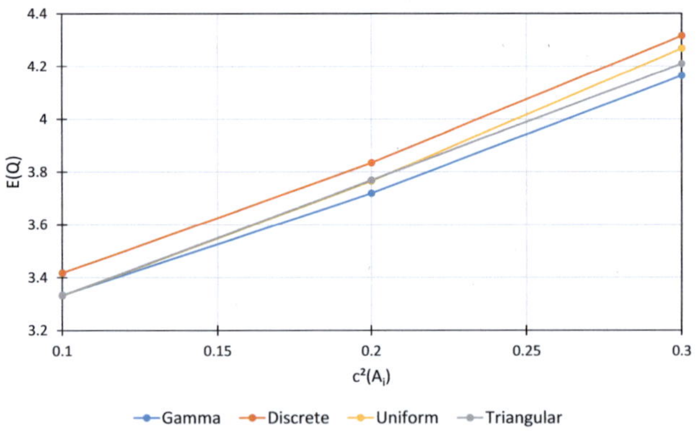

Figure 7.1: Impact of the interarrival time variability $c^2(A_i)$ with different distribution types on the total queue length $E(Q)$

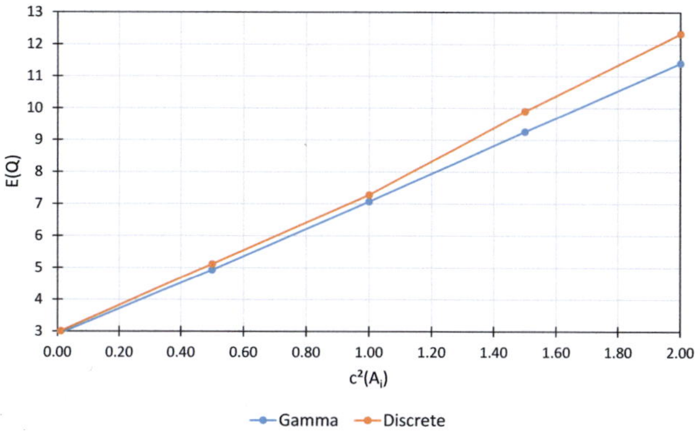

Figure 7.2: Impact of a higher interarrival time variability $c^2(A_i)$ with different distribution types on the total queue length $E(Q)$

With increasing variability of the interarrival time, a linear increase in the total number of waiting customers in the system can be observed. The type of distribution of the interarrival time has almost no influence on the total queue length with the same expected value and variability (Figure 7.1). With a gamma distributed interarrival time, the total number of waiting customers in the system is slightly lower. A discrete distribution with ten possible values, where the probabilities are randomly generated based on the given expected value and variability, leads to a slightly higher total queue length.

For a higher variability of the interarrival time, this fact is clearly recognizable (Figure 7.2). The difference in $E(Q)$ for gamma and discrete distributed interarrival times can be explained by the total variability of the interarrival time. The variability of the merged arrival stream for gamma distributed interarrival times of the single streams is below the variability for discrete distributed interarrival times. The variability of the total interarrival time does not increase to the same extent as the interarrival time of the individual queues. With a gamma distribution and $c^2(A_i) < 1$ the variability of the combined stream is higher than the variability of the single streams.

With $c^2(A_i) > 1$, $c^2(A)$ is lower compared to $c^2(A_i)$. With $c^2(A_i) = 1$ and thus exponentially distributed interarrival times $c^2(A_i) = c^2(A)$ applies. Due to the independency based on the exponential distribution, the combined distribution of the interarrival time is also exponentially distributed and the variability is 1. With discretely distributed interarrival time, the independency is not given and the variability of the total interarrival time increases significantly at $c^2(A_i) > 1$. This correlation of the variabilities of joined flows can be described by the stochastic merge of (Furmans 2004).

In the investigation of the impact of the distribution of arrivals over the queues with different interarrival time variabilities, a correlation between the total queue length and the variability of the joined stream can also be observed. With increasing difference of the interarrival times over the queues and $c^2(A_i) < 1$ the total queue length decreases (Figure 7.3). With $c^2(A_i) > 1$ on the other hand, the total number of waiting customers in the system increases and with $c^2(A_i) = 1$ the distribution of arrivals over the queues has no influence on the number of customers in the system. This behaviour is again due to the variability of the total interarrival time.

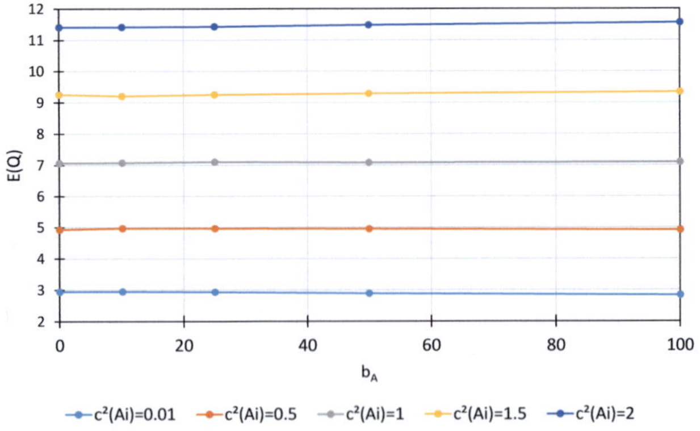

Figure 7.3: Impact of the distribution of arrivals over the queues measured by the gradient b_A with different interarrival time variabilities $c^2(A_i)$ on the total queue length $E(Q)$

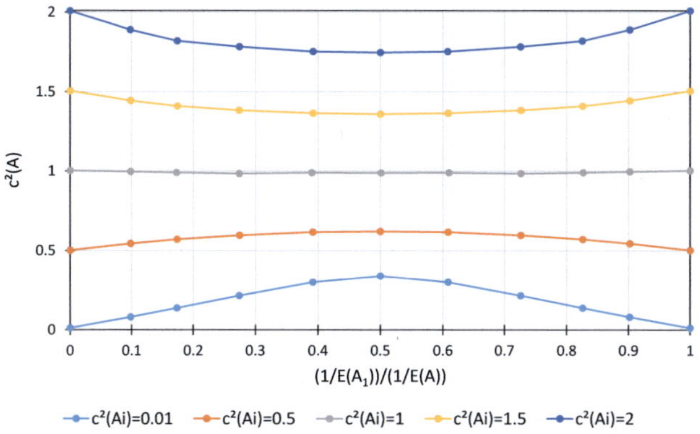

Figure 7.4: Impact of the distribution of arrivals over the queues measured by the ratio $(1/E(A_1))/(1/E(A))$ with different interarrival time variabilities $c^2(A_i)$ on the total interarrival time variability $c^2(A)$

The behaviour of the total interarrival time variability for two queues and the variation of the percentage of the throughput of queue 1 in relation to the total throughput is shown in Figure 7.4 for the different $c^2(A_i)$. At $c^2(A_i) < 1$, the variability of the total flow is higher with uniformly distributed throughput across the queues. The total interarrival time variability increases by 327% when the ratio increases from 0.1 to 0.5 and $c^2(A_i) = 0.01$, and by only 14% with $c^2(A_i) = 0.5$. If $c^2(A_i) > 1$ applies, $c^2(A)$ is smaller with the same distribution of arrivals over the queues. With a variability of 1 of the interarrival time distributions of the queues $c^2(A) = c^2(A_i) = 1$ applies due to the exponential distribution independent of the distribution of the arrivals over the queues.

Considering the impact of the number of queues with different interarrival time variabilities on the total queue length, there are also different behaviours related to the variability of the interarrival times of the queues (Figure 7.5). The total number of waiting customers in the system increases with an increase of N and $c^2(A_i) < 1$ and with $c^2(A_i) > 1$ it decreases.

For $c^2(A_i) = 1$, the total number of waiting customers in the system is independent of the number of queues. Again, this can be explained by the variability of the merged stream (Figure 7.6). For $N \to \infty$ the total interarrival time variability converges to 1 independent of $c^2(A_i)$.

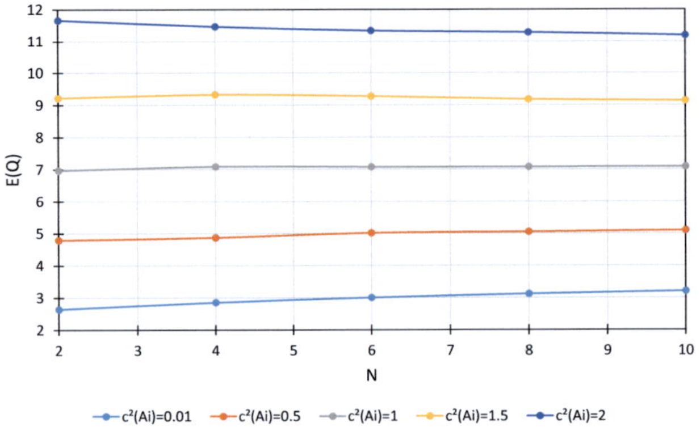

Figure 7.5: Impact of the number of queues N with different interarrival time variabilities $c^2(A_i)$ on the total queue length $E(Q)$

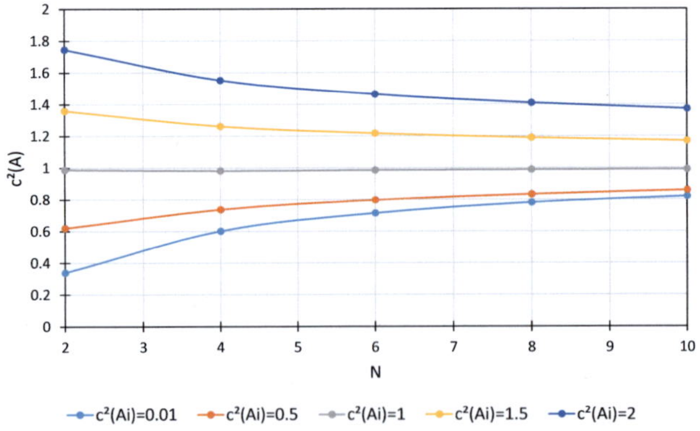

Figure 7.6: Impact of the number of queues N with different interarrival time variabilities $c^2(A_i)$ on the total interarrival time variability $c^2(A)$

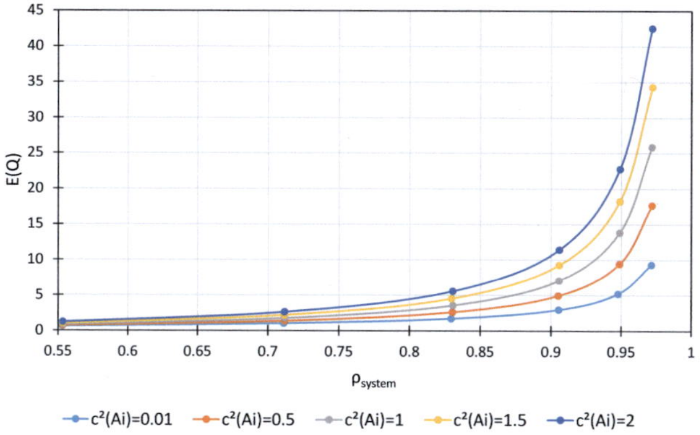

Figure 7.7: Impact of the utilization of the system ρ_{system} with different interarrival time variabilities $c^2(A_i)$ on the total queue length $E(Q)$

By increasing the expected value of the interarrival time, the utilization of the system increases. As typically observed in queuing systems, there is an exponential increase of the total queue length with increasing utilization (Figure 7.7). The increase is more steeper with higher variability of the interarrival time.

If the capacity of the queues is reduced, the probability of rejection increases (Figure 7.8). If the variability of the interarrival time is higher, p_{reject} increases earlier and faster. With $c^2(A_i) = 2$ more than 30% are rejected with $K_i = 1$ despite a utilization under 1. With a very low variability ($c^2(A_i) = 0.01$) only 4% are rejected even with $K_i = 1$.

The results of the evaluation of the system characteristics in relation to the processing time are shown in the figures 7.9-7.16. Again, only the expected value of the total number of waiting customers in the system is considered, since the expected value of the waiting time and sojourn time behave in the same way with respect to the parameter settings considered.

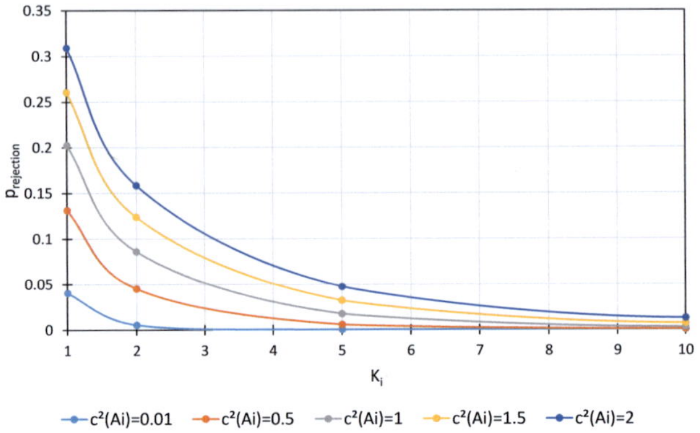

Figure 7.8: Impact of the queue capacity K_i with different interarrival time variabilities $c^2(A_i)$ on the total rejection probability p_{reject}

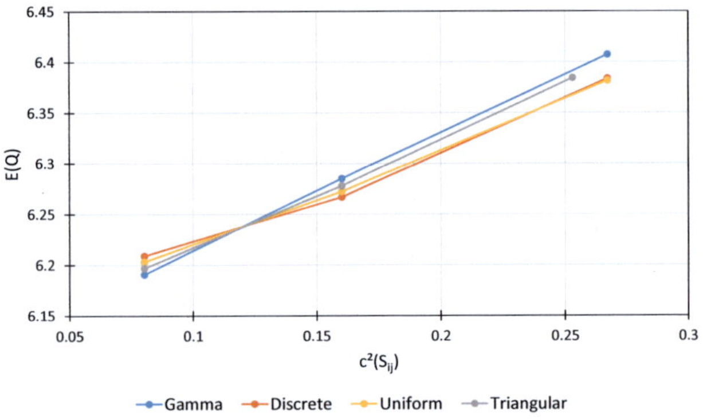

Figure 7.9: Impact of the processing time variability $c^2(S_{ij})$ with different distribution types on the total queue length $E(Q)$

By increasing the variability of the processing time, the total number of waiting customers increases (Figure 7.9). The distribution of the processing time has almost no influence on the total queue length in case of the same $E(S)$ and $c^2(S_{ij})$.

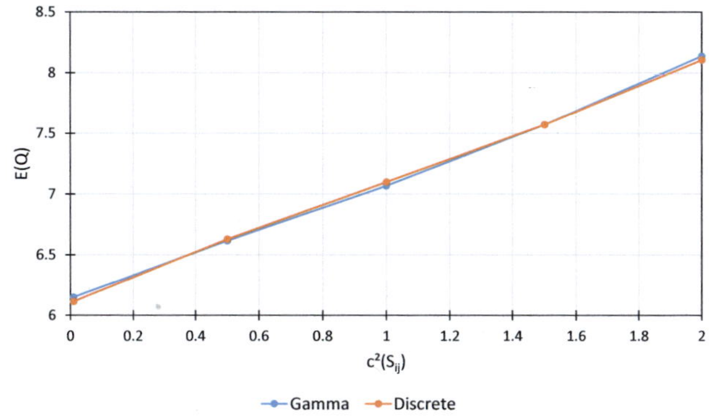

Figure 7.10: Impact of a higher processing time variability $c^2(S_{ij})$ with different distribution types on the total queue length $E(Q)$

Also when considering higher processing time variabilities, the total queue length is approximately the same for gamma distributed and discrete distributed processing times (Figure 7.10). However, the influence of the variability of the processing time is much smaller than the influence of the variability of the interarrival time. A reduction of the variability $c^2(S)$ from 1 to 0.01 reduces the total queue length by 13%. A reduction of $c^2(A_i)$ from 1 to 0.01 reduces the total queue length by 58%.

Observing the distribution of the processing time across the queues and a processing time independent service rule, the total queue length increases with increasing gradient of the $E(S)$ across the queues (Figure 7.11). The increase is almost independent of the variability of the processing time.

If the selected service rule is dependent on the expected value of the processing time, the behaviour of the total number of waiting customers in the system changes with a variation of b_{S_i}. Thus, with service rule 7 the total queue length decreases with increasing gradient (Figure 7.12) and with service rule 8 the total number of waiting customers increases even more than with a processing time independent service rule (Figure 7.13).

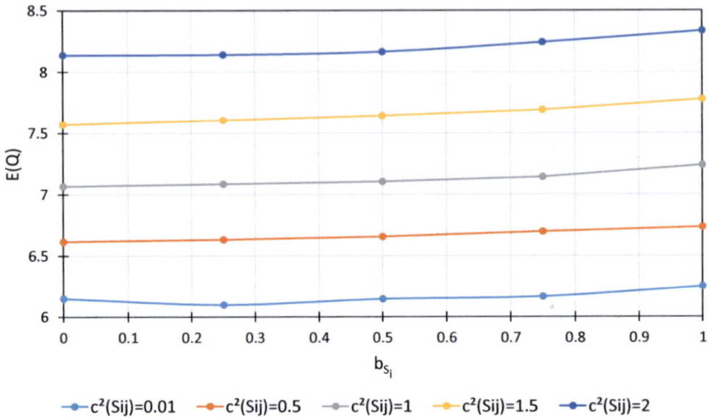

Figure 7.11: Impact of the distribution of the processing time over the queues measured by the gradient b_{S_i} with different processing time variabilities $c^2(S_{ij})$ and processing time independent service rule (SR2) on the total queue length $E(Q)$

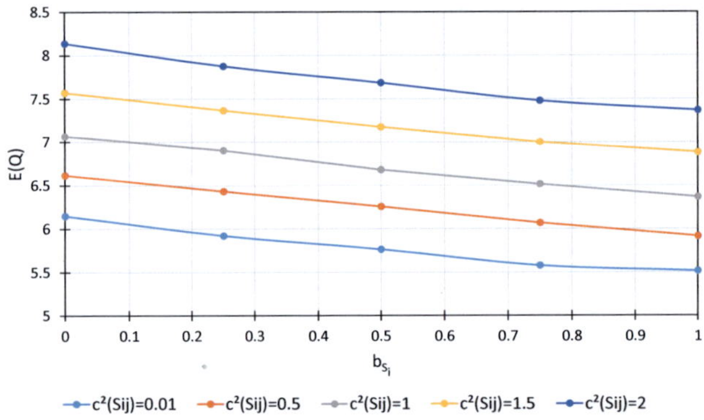

Figure 7.12: Impact of the distribution of the processing time over the queues measured by the gradient b_{S_i} with different processing time variabilities $c^2(S_{ij})$ and processing time dependent service rule (SR7) on the total queue length $E(Q)$

When the gradient of the expected value across the sinks is increased, the same characteristic is observed as with the gradient of the expected value across the queues and independent service rule (Figure 7.14). This behaviour is independent of the service rule when considering b_{S_j}.

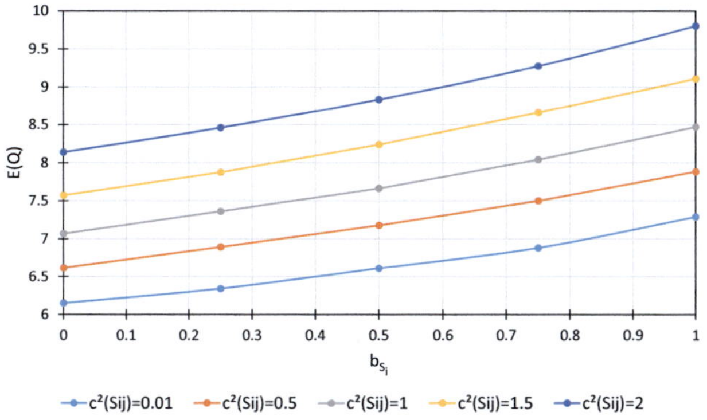

Figure 7.13: Impact of the distribution of the processing time over the queues measured by the gradient b_{S_i} with different processing time variabilities $c^2(S_{ij})$ and processing time dependent service rule (SR8) on the total queue length $E(Q)$

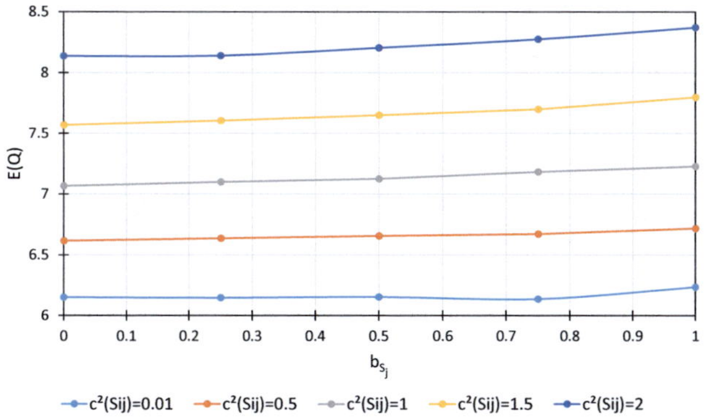

Figure 7.14: Impact of the distribution of the processing time over the sinks measured by the gradient b_{S_j} with different processing time variabilities $c^2(S_{ij})$ on the total queue length $E(Q)$

An increase in the utilization ρ_{system} can also be achieved by increasing $E(S)$. Again, the variability of the processing time affects the queue length less than the interarrival time variability (Figure 7.15). Even if K_i is reduced, the influence of $c^2(S_{ij})$ on the rejection probability is lower (Figure 7.16).

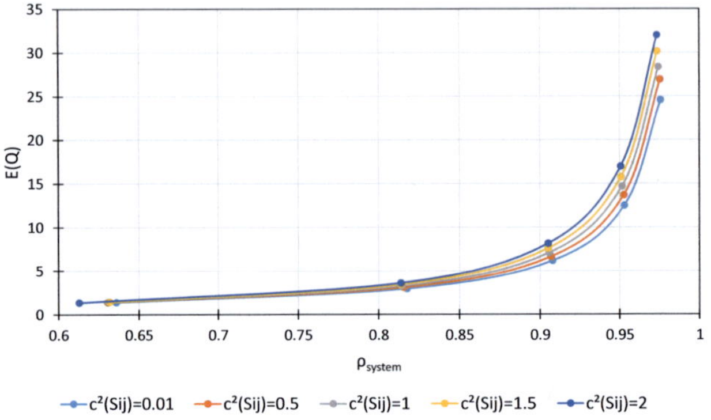

Figure 7.15: Impact of the utilization of the system ρ_{system} with different processing time variabilities $c^2(S_{ij})$ on the total queue length $E(Q)$

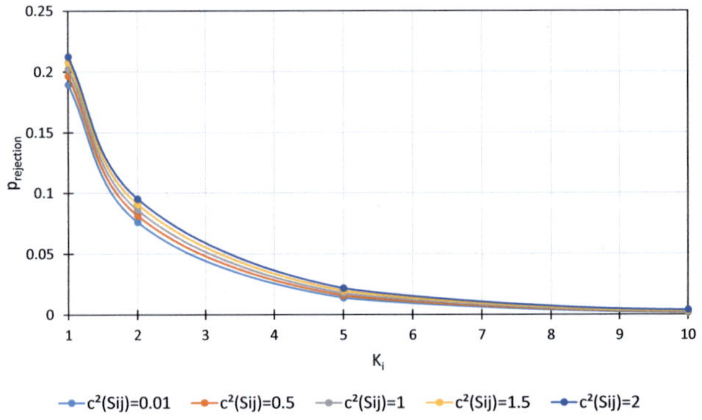

Figure 7.16: Impact of the queue capacity K_i with different processing time variabilities $c^2(S_{ij})$ on the total rejection probability p_{reject}

Figures 7.17-7.24 shows the results of the evaluation of the system characteristics in relation to the switching time. The behaviour of the system when varying the parameters relative to the switching time is equal to the behaviour when varying the equivalent processing time parameters.

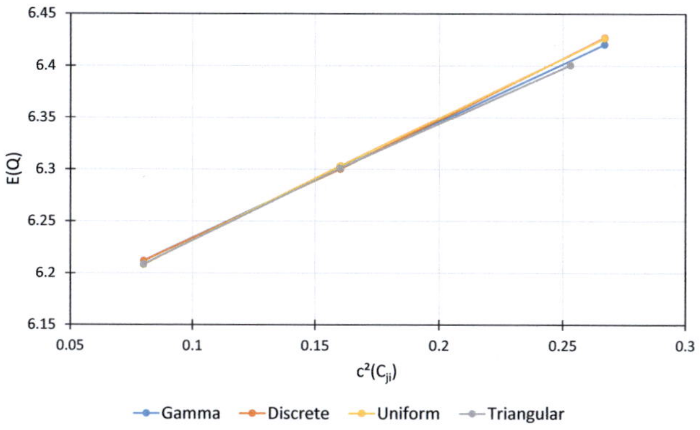

Figure 7.17: Impact of the switching time variability $c^2(C_{ji})$ with different distribution types on the total queue length $E(Q)$

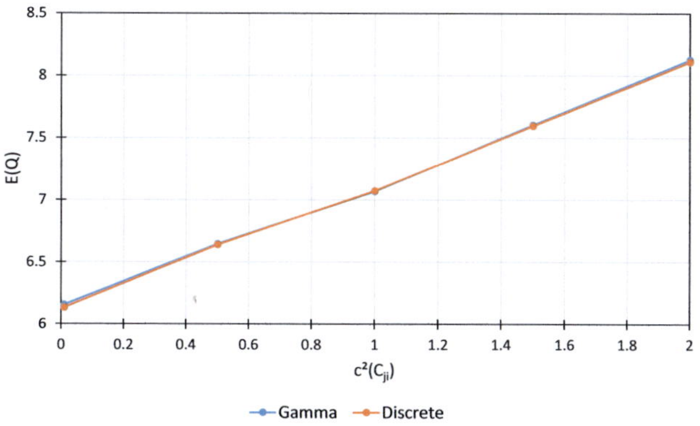

Figure 7.18: Impact of a higher switching time variability $c^2(C_{ji})$ with different distribution types on the total queue length $E(Q)$

The switching time dependent service rule 6 reduces $E(Q)$ with increasing b_{C_i} (Figure 7.20). If BV_2 is additionally set to 1 so that there is no switching time when switching from a sink to the queue with the same number, $E(Q)$ is significantly reduced (Figure 7.21). The increase of b_{C_i} initially leads to an increase in $E(Q)$. However, with a higher b_{C_i}, $E(Q)$ decreases again.

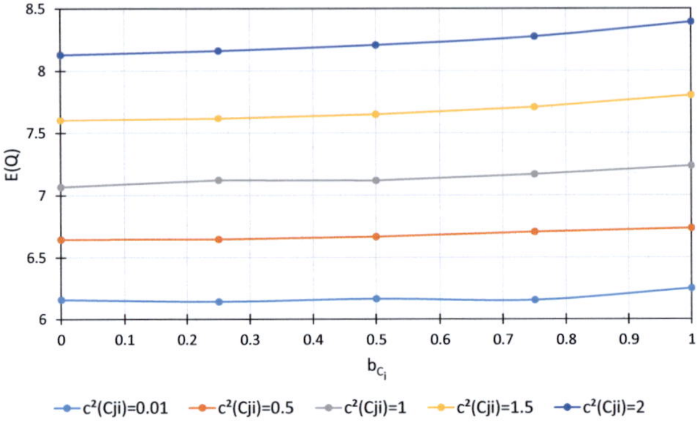

Figure 7.19: Impact of the distribution of the switching time over the sinks measured by the gradient b_{C_j} with different switching time variabilities $c^2(C_{ji})$ and switching time independent service rule (SR2) on the total queue length $E(Q)$

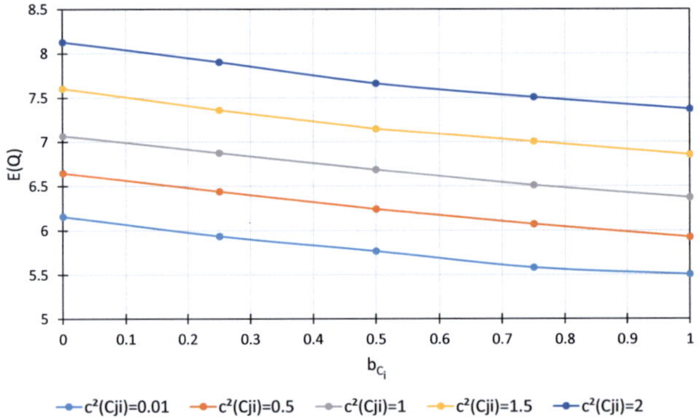

Figure 7.20: Impact of the distribution of the switching time over the sinks measured by the gradient b_{C_j} with different switching time variabilities $c^2(C_{ji})$ and switching time dependent service rule (SR7) on the total queue length $E(Q)$

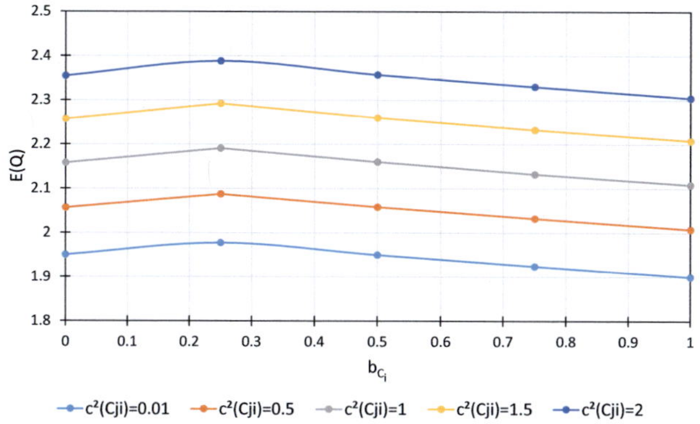

Figure 7.21: Impact of the distribution of the switching time over the sinks measured by the gradient b_{C_j} with different switching time variabilities $c^2(S_{ji})$ and switching time dependent service rule (SR8) on the total queue length $E(Q)$

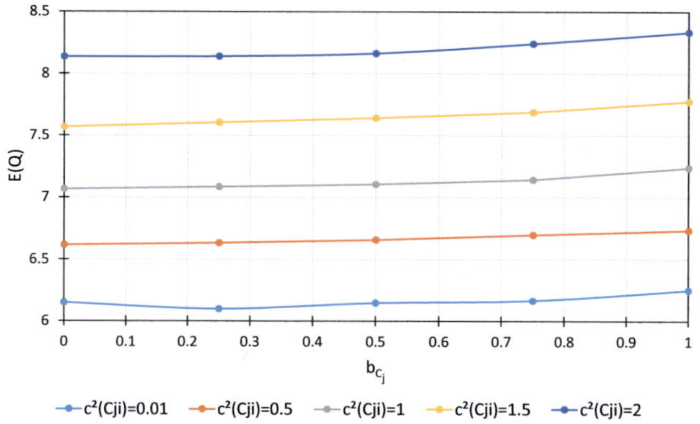

Figure 7.22: Impact of the distribution of the switching time over the sinks measured by the gradient b_{C_i} with different switching time variabilities $c^2(C_{ji})$ on the total queue length $E(Q)$

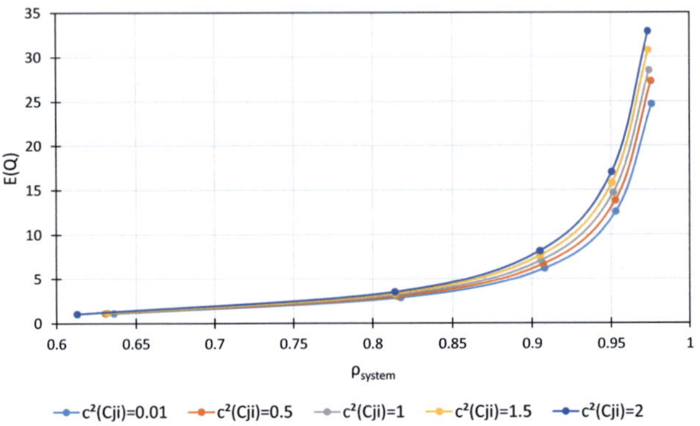

Figure 7.23: Impact of the utilization of the system ρ_{system} with different switching time variabilities $c^2(C_{ji})$ on the total queue length $E(Q)$

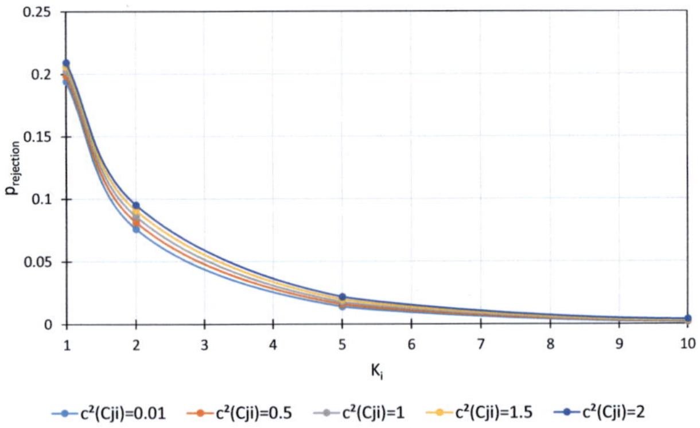

Figure 7.24: Impact of the queue capacity K_i with different switching time variabilities $c^2(C_{ji})$ on the total rejection probability p_{reject}

7.2.2 Evaluation of the Service Rules

Based on the analyses of the system characteristics, the various service rules are considered individually and evaluated on the basis of further service rule specific analyses. To evaluate the service rules, service rules 1-10 (abbreviated SR1-SR10) of rule category 1 are analysed first. Service rules 1-10 consist of rule types 1.1-1.10 from rule category 1 and no rule type from rule category 2. By considering service rule 1-10 the focus is on the question *which queue is selected next* (rule category 1). The results are shown in the figures 7.25-7.28. As the sojourn time distribution corresponds to the waiting time distribution, the results are not presented in relation to the sojourn time. The service rules 1 and 10 are not shown in the diagrams, since for these rules the utilization of the system is 1 and the performance parameters are therefore significantly higher compared to the other rules.

In the parameter settings considered, for service rule 1 $\rho_{system} = 1$ applies, due to the time gaps in which the server stands still even though customers are waiting. However, $\rho_{server} \approx \tilde{\rho}$ and there are no rejections. This service rule enables the system to stabilize at a utilization of 1. In service rule 1 the time where the server is idle even if a customer is waiting in a queue corresponds to the idle time of the system with service rules where the server cannot be empty when a customer is in a queue. In the case of an unequal distribution of the arrivals over the queues, service rule 1 prioritizes the lower arrival stream due to the service of the customers of the queues in a fixed order. The server may then be idle although many customers wait in the queue with the lower mean interarrival time. This leads to a high queue length and waiting time of the customers in the queue with more frequent arrivals and can lead to an overload of the system. Overall, service rule 1 leads to a higher number of waiting customers in the system, longer waiting times and longer sojourn times on average as well as for the 95%-quantile.

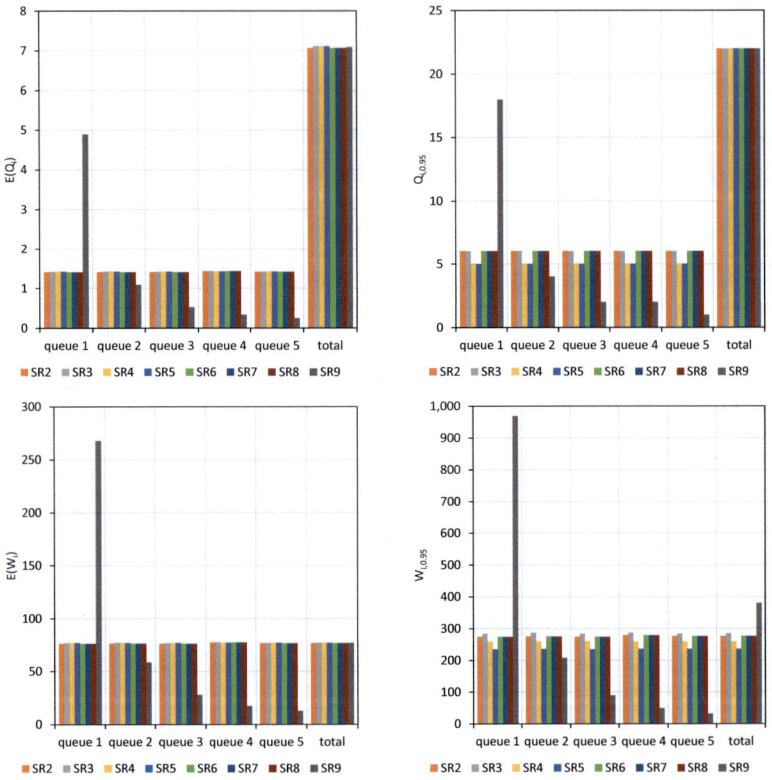

Figure 7.25: Expected value and 95%-quantile of the number of waiting customers and waiting time in the system per queue and total for the basic setting

With service rule 2, empty queues are skipped and no idle server occur when a customer is in the system. The service rule is independent of the interarrival, processing and switching time, as well as the current system state. Accordingly, no special characteristics are used to improve the performance parameters. With this service rule, an unequal distribution of the arrivals over the queues will prioritize the queue with less arrivals. This leads to a higher expected value and 95%-quantile of the queue length and waiting time of the queue with the lower mean interarrival time (see Figure 7.26).

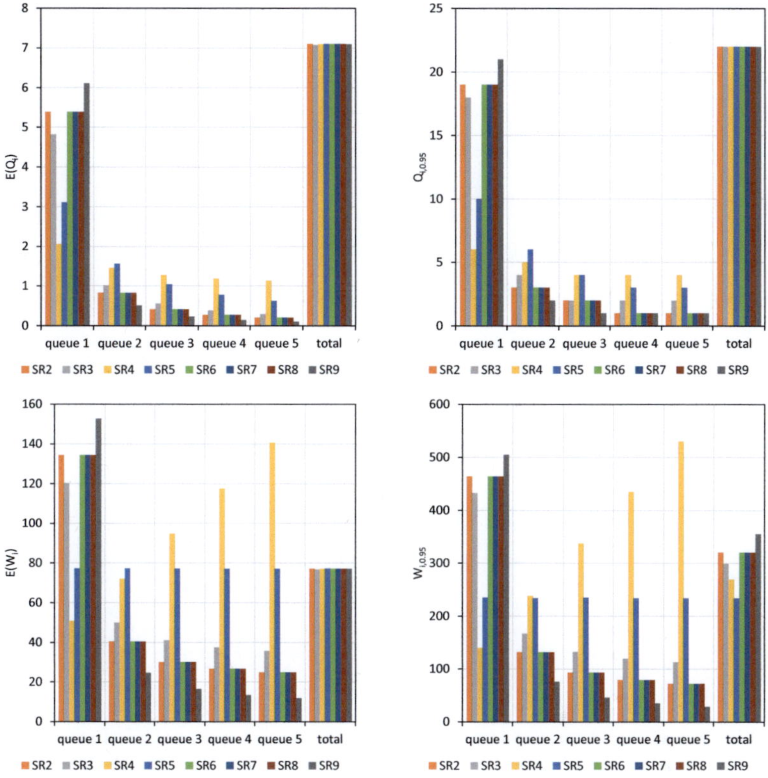

Figure 7.26: Expected value and 95%-quantile of the number of waiting customers and waiting time in the system per queue and total in case of an unequal distribution of arrivals over the queues

Service rule 3 behaves similarly to service rule 2. The random selection based on a probability distribution, however, increases the variability of the system and thus the variability of the performance parameter distributions. This can be seen in the 95%-quantile of the waiting time distribution in Figure 7.25. In case of unequal interarrival times, the uniform distribution for selecting the next queue leads to a prioritisation of the queue with less arrivas. A distribution weighted according to the throughput resolve this.

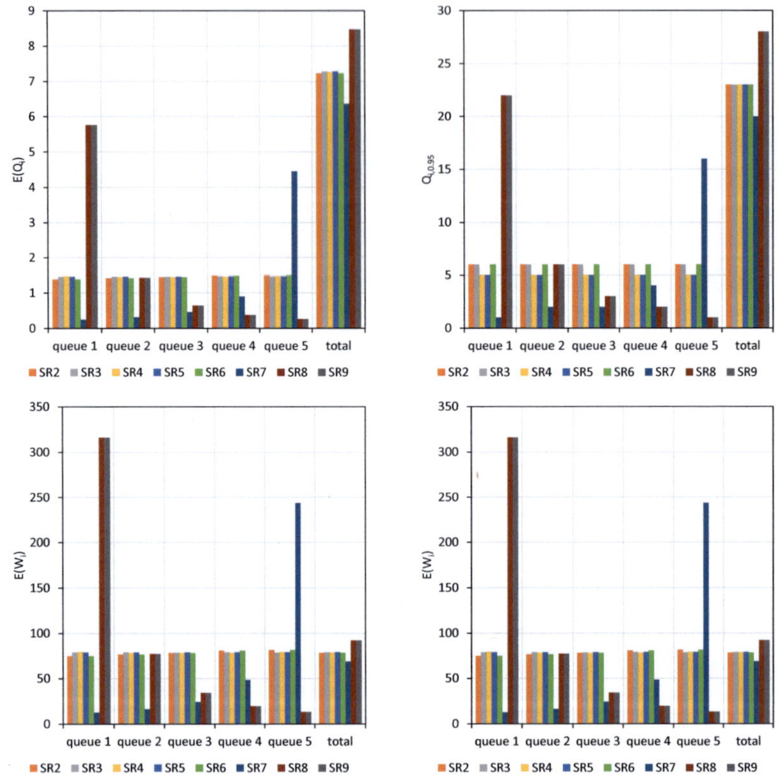

Figure 7.27: Expected value and 95%-quantile of the number of waiting customers and waiting time in the system per queue and total in case of an unequal distribution of the processing time over the queues

Service rule 4 ensures a uniform service for the customers of the queues by selecting the maximum queue as the next queue to be selected. The uniformity is also reflected in the variability of the distribution of the number of waiting customers in the system per queue. Compared to service rule 2, the variability in the basic setting is reduced from 2.45 to 1.35 by 44.9%. This can also have a positive effect on the expected value and the 95%-quantile of the distribution of the number of waiting customers in the system.

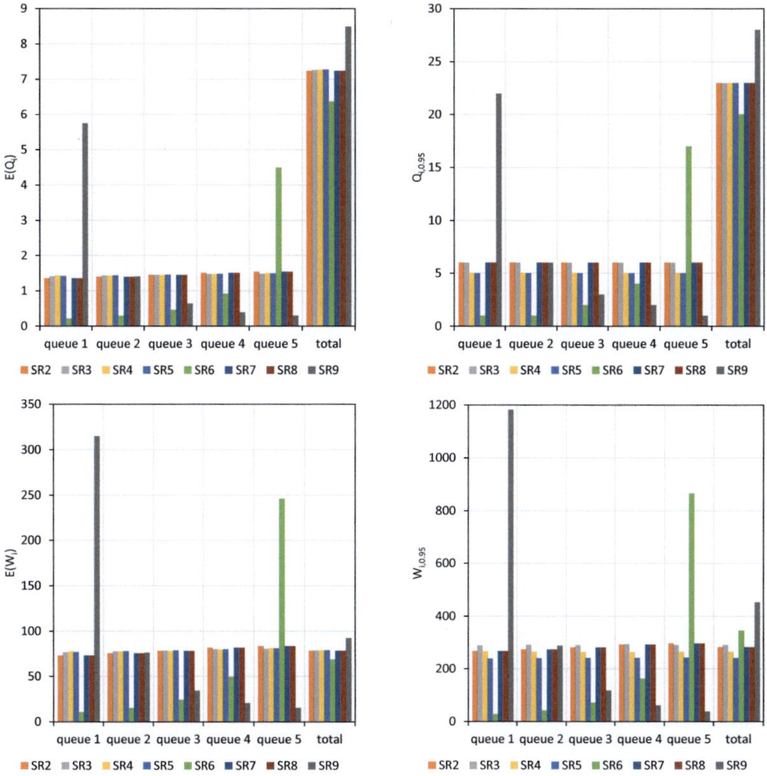

Figure 7.28: Expected value and 95%-quantile of the number of waiting customers and waiting time in the system per queue and total in case of an unequal distribution of the switching time over the queues

A uniform distribution of arrivals over the queues also reduces the variability of the waiting time distribution compared to service rule 2. If the distribution is unequal, the queue with a higher arrival rate is prioritized, which leads to a higher average waiting time of the queue with the lower arrival stream (see Figure 7.26). The 95%-quantile of the total waiting time of a customer in the system is also reduced. Altogether, service rule 4 can ensure a uniform service of the queues.

Service rule 5 balances the variation in waiting time by selecting the queue with the maximum waiting time. Compared to service rule 2, the variability of the distribution of the waiting time of a customer in the system per queue is reduced from 1.80 to 1.35 by 40.8% in the basic setting. Figure 7.25 shows the effect of the reduction on the 95%-quantile of the waiting time of a customer in the system per queue and total. The quantile reduces from 276 to 236 time units by 14.5% related to service rule 2. In the case of unequal interarrival times over the queues, service rule 5 compensates for the inequality so that the mean waiting time of a customer is the same across all queues. The compensation reduces the 95%-quantile of the total waiting time of a customer in the system by 26.9% (see Figure 7.26). Service rule 5 enables a balanced average waiting time and thus a balanced sojourn time over the queues.

If the expected value of the switching time is uniform across all queues, service rule 6 corresponds to service rule 2 due to the *additional rule for ambiguous decision* (see Chapter 4). In case of an unequal distribution of the switching time over the queues, service rule 6 switches to the non-empty queue with the lowest mean switching time. By time-minimizing switching, the mean switching time can be reduced overall and thus the utilization, the mean queue length, the mean waiting time and the mean sojourn time can be reduced. In the parameter setting considered, $E(Q)$ is reduced by 11.9% and $E(W)$ by 12.1% based on service rule 2 (see Figure 7.27). However, depending on the switching time matrix, service rule 6 may discriminate a queue with high switching times. Accordingly, the expected value and the 95%-quantile of the queue length and waiting time for this queue increase. Figure 7.28 shows the increase at queue 5. If there is no switching time from sink to queue with the same number and otherwise a uniform distribution of the switching time over the queues, service rule 6 can be used to improve the performance parameters without discrimination. The symmetry in the mean switching time matrix allows a reduction of the mean switching time without prioritization. In the parameter setting under consideration, the condition *no switching time from sink to queue with the same number* leads to a reduction of the average total number of customers in the system by 73.1% and the average total waiting time of a customer in the system by 74.0%. The respective 95%-quantiles are reduced by 76.0% and 75.6% respectively.

In the same way as with service rule 6, service rule 7 corresponds to service rule 2 in the case of uniform distribution of the processing time over the queues due to the additional rule for ambiguous decision. If the expected value of the processing time is unequal distributed over all queues, an improvement of the performance parameters can be achieved by *shortest job first*. The queue with the lowest mean processing time is prioritized. This leads to a reduction of the total mean queue length and the total mean waiting time of a customer in the system, but only due to the disadvantage of the queue with the higher mean processing time. In the parameter setting considered, $E(Q)$ is reduced by 11.9% and $E(W)$ by 12.1%. $E(Q_5)$ and $E(W_5)$ increases by 196.2% and 198.6%, respectively (see Figure 7.27). In comparison to service rule 4 and 5, the 95%-quantile of the queue length and waiting time increases for the parameter setting considered.

Like service rule 7, service rule 8 behaves in the case of uniform distribution of the processing time over the queues identical to service rule 2. In case of an unequal distribution of the processing time over the queues, the expected value of the number of customers and the expected value of the waiting time as well as the corresponding 95%-quantile increases.

With service rule 9 there is an explicit prioritization of the queues. Accordingly, the performance parameters for high-priority queues are better than those for less prioritized queues. Overall, this service rule leads to higher 95%-quantiles in queue length and wait time (see Figure 7.25-7.28). In case of an unequal distribution of arrivals over the queues and prioritization of the queue with a higher arrival rate, more balanced queue lengths can be achieved by service rule 9. If the expected value of the processing time is unequal distributed over all queues, service rule 9 can be used to represent service rule 7 or 8 by prioritizing the queues according to the mean processing time. In Figure 7.27, service rule 9 corresponds to service rule 8, since the queues were prioritized according to the descending mean processing time.

With service rule 10, the utilization of the system is equal to 1 and customers of less prioritized queues are rejected. Due to the premise that no customer may wait for a customer of lower priority, long idle times of the server occur even though customers wait in the system at lower prioritized queues. In the basic setting considered, the server is only utilized to 25.4% while the estimated utilization is 90.9% (Section 6.2). This low utilization of the server leads to an overall overload of the system. With low variabilities of processing and switching time as well as the prioritization of a queue with very few arrivals, a load below 1 can be achieved for service rule 10. However, also in this case the expected value of the number of customers in the system in total and the expected value of the waiting time of a customer in the system in total as well as the corresponding 95%-quantile is larger compared to the other service rules. The mean queue length and the mean waiting time of the high-priority queue, on the other hand, is approximately 0.

When using service rules that refer to rule types of rule category 2, customers of the same queue are served one after the other. This is especially useful if there is no switching time between two customers from one queue. This can be modelled using a deterministic transition and the condition *no switching time from sink to queue with the same number*. The figures 7.29-7.32 show the results of the investigations of the service rules of rule category 2. The analysis is limited to the case of no switching time from sink to queue with the same number and deterministic transitions which is the most reasonable use of the service rules of rule category 2.

In order to apply a service rule with rule types 2.2, 2.3, 2.5 or 2.6, the additional parameters for the corresponding rule type must first be determined. For the considered parameter setting the following parameters are determined by parameter variation and performance evaluation: $TW_i = 14$, $MN_i = 2$, $LV_i^1 = 66$ and $LV_i^2 = 3100$.

Since the investigations of the service rules related to rule category 2 cannot be carried out independently of the rule types from rule category 1, useful combinations must be defined. From the respective results of the investigation of certain combinations, further combinations can be specified for which useful results can be expected.

In a first step the basic setting (see appendix C.2 in Table C.2) is adapted so that there is no switching time from sink to queue with the same number and there are deterministic transitions. The service rules with rule types 2.1-2.6 in combination with rule type 1.2 are analysed. The results are shown in Figure 7.29. It is obvious that with service rules where the number of operations of a queue is determined by an additional parameter (service rule 22, 32, 52 and 62) a higher mean number of waiting customers and a higher mean waiting time of a customer in the system per queue and in total is reached. The 95%-quantiles are higher than with the service rules with exhaustive (service rule 12) or gated (service rule 42) service per queue. This is caused by the idle times of the server while customers are waiting, which occurs with these service rules. For the service rules with exhaustive or gated service per queue there are no idle times of the server when a customer is in the system. This leads to better performance parameters.

In a second step rule types 2.2, 2.3, 2.5 or 2.6 with rule type 2.1 (*exhaustive*) are combined to prevent idle times of the server even though customers are waiting. This corresponds to the service rules 72, 82, 122 and 162. Furthermore, the service rule 12 (*exhaustive*) and service rule 42 (*gated*), which are also rules with no idle time of this kind, are considered.

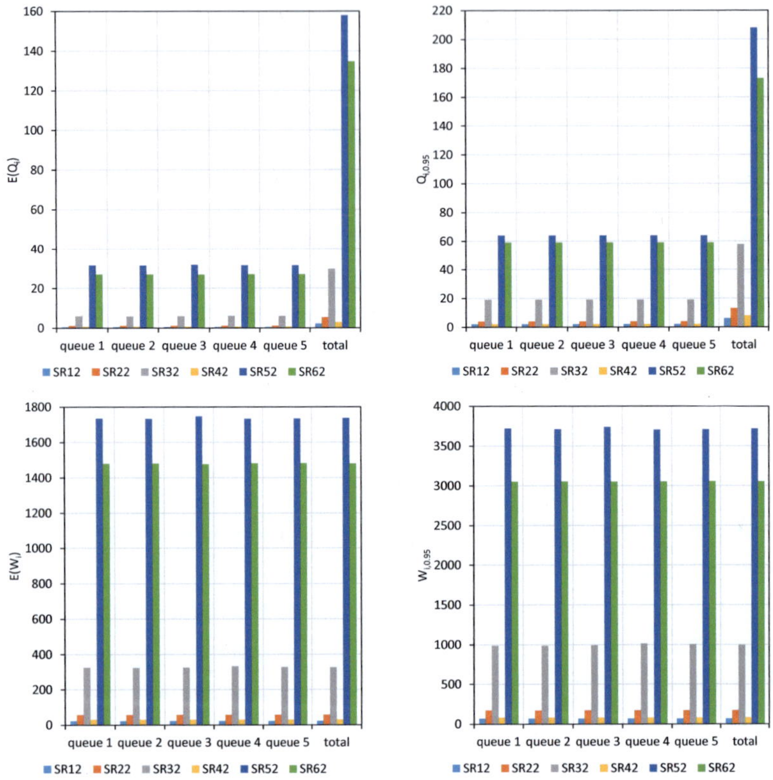

Figure 7.29: Expected value and 95%-quantile of the number of waiting customers and waiting time in the system per queue and total in case of no switching time from sink to queue with the same number and deterministic transitions

Figure 7.30 shows the comparison of the service rules 12, 42, 72, 82, 122 and 162. The mean number of waiting customers and the mean waiting time per queue and in total of the service rules 72, 82, 122 and 162 is significantly lower compared to the service rule without exhaustive service. Service rule 122 and 162 which are based on limited values of the other queues behave identical to service rule 12, since before the limited value occurs the current queue is empty and is switched. The limited value is therefore not necessary.

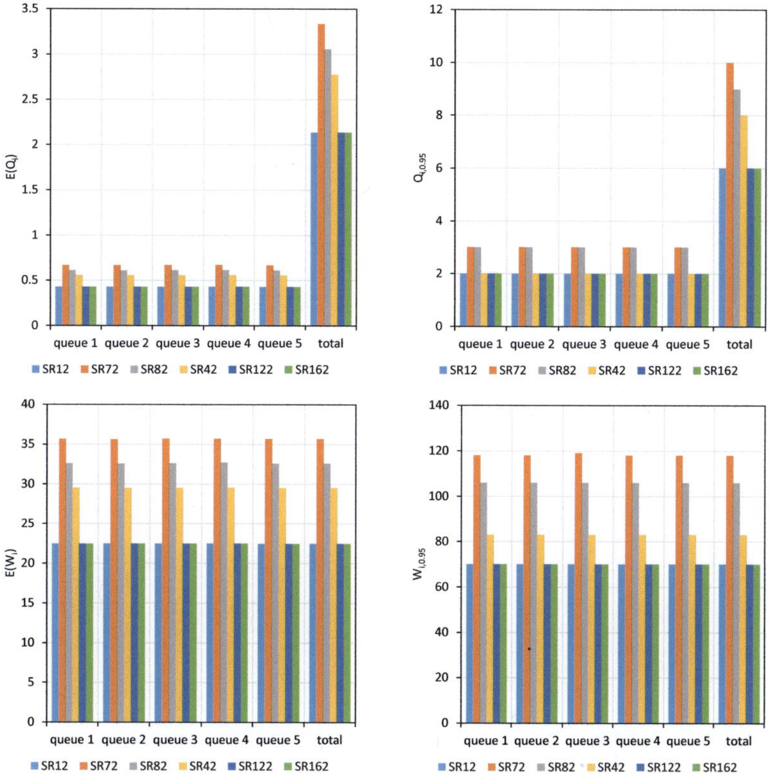

Figure 7.30: Expected value and 95%-quantile of the number of waiting customers and waiting time in the system per queue and total in case of no switching time from sink to queue with the same number and deterministic transitions for service rules with exhaustive or gated service per queue

The expected value and the 95%-quantile of the waiting customers and the waiting time of a customer in the system per queue and in total are slightly higher for service rule 72 and 82 than for service rule 12. Due to the fixed time window or the fixed number of customers to be served, the system switches although customers are still waiting in the last selected queue. This leads to higher switching times, which negatively influence the performance parameters.

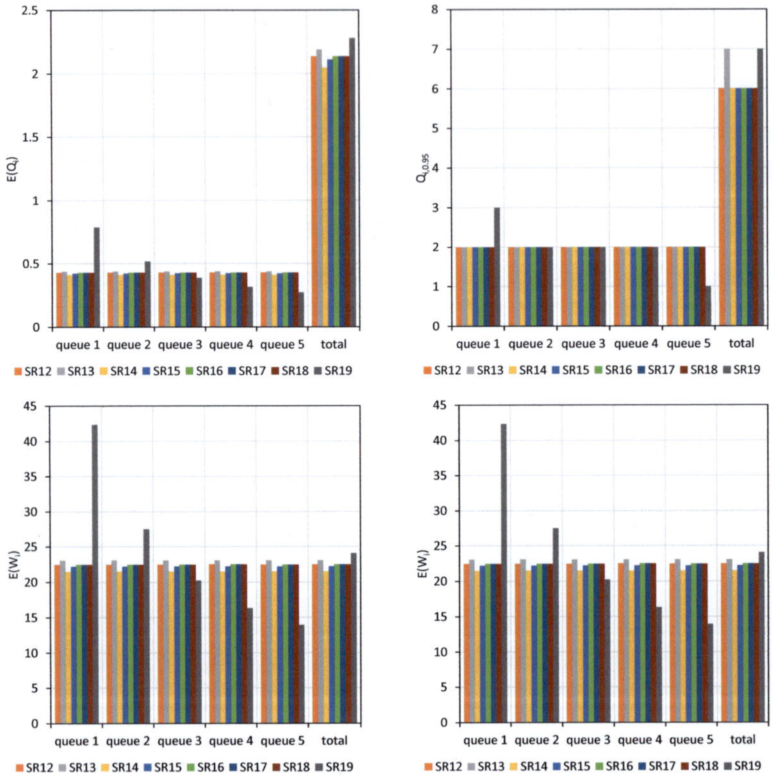

Figure 7.31: Expected value and 95%-quantile of the number of waiting customers and waiting time in the system per queue and total in case of no switching time from sink to queue with the same number and deterministic transitions for service rules with exhaustive service per queue

Since the service rules with *exhaustive* (rule type 2.1) and *gated* (rule type 2.4) lead to better performance parameters, in a further step these rule types are combined with the rule types 1.1-1.10 of rule category 1 (see Figure 7.31 and 7.32). The service rules with the rule types 1.1 and 1.10 are not shown due to the overloaded system.

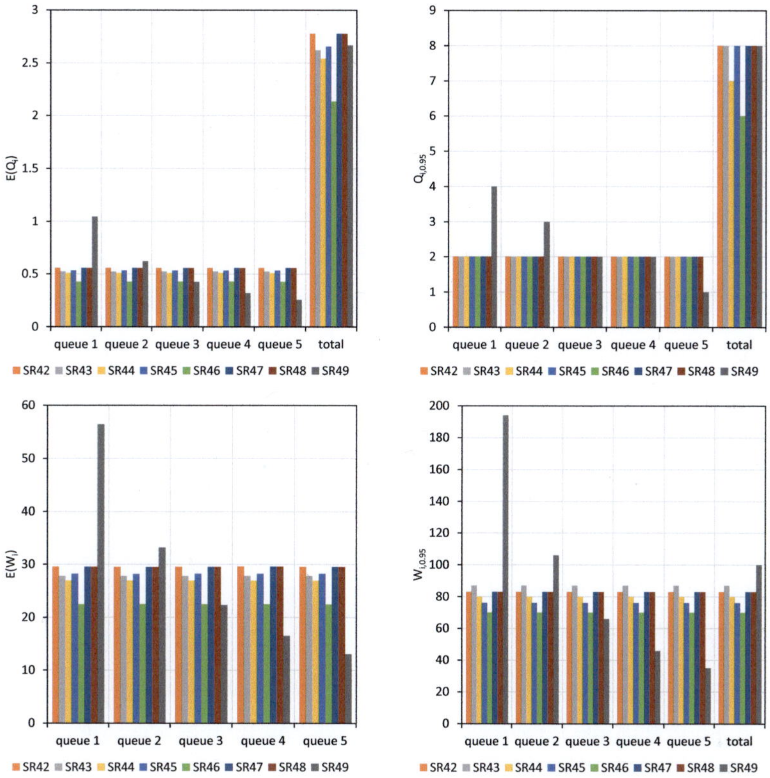

Figure 7.32: Expected value and 95%-quantile of the number of waiting customers and waiting time in the system per queue and total in case of no switching time from sink to queue with the same number and deterministic transitions for service rules with gated service per queue

Independent of the rule type selected of rule category 1, it is shown that the combination with exhaustive service leads to better results than gated service in the parameter setting considered. Since with the service rules with exhaustive service customers of the same queue are served even longer in a row, the switching occurs even less frequently and thus the average switching time with exhaustive service is reduced compared to gated service.

Only for rule type 1.6 (*non-empty queue with minimum switching time*) the service rule with exhaustive (service rule 16) and gated service (service rule 46) leads to the same performance parameters. This can be explained by the fact that this service rule selects the queue with the lowest switching time. With gated service and further customers in the currently selected queue, the next time a queue is selected, the system switches again to the same queue. Thus service rule 46 behaves identically to service rule 16 with the parameter setting *no switching time between two customers from one queue*.

Using service rule 14, the analysis shows a reduction of the mean number of waiting customers and the mean waiting time of a customer in the system per queue and in total. When the queue changes, the queue with the maximum number of customers is selected. This means that with exhaustive service, a large number of customers can be served from one queue. As a result, the switching time is reduced and the utilization, average queue length, waiting time and sojourn time of a customer in the system per queue and in total are reduced.

Using service rule 15, the waiting time can be balanced equivalent to service rule 5 and thus the 95%-quantile of the waiting time of a customer in the system per queue and total can be reduced.

Service rule 12, 16, 17 and 18 behave identically due to the uniform distribution of the switching and processing time over the queues. The behaviour is the same as for service rule 6, due to the condition *no switching time from sink to queue with the same number* this service rule always switches to the same queue as long as a customer is in this queue and then switches to the next queue in the order. If the processing time is distributed unequally over the queues, service rule 17 reduces the performance parameters compared to service rule 6.

7.2.3 Recommendations of Service Rules

The evaluation of the service rules can be used to identify the influencing parameters that affect the selection of a service rule. The 33 input parameters in Table 7.1 can therefore be reduced to five parameters (Table 7.4.)

Category	Parameters	Notation	Specification
Transition	Distribution of the probability across the sinks	$D_{\hat{p}}$	uniform, unequal, deterministic
Interarrival times	Distribution of the expected value across the queues	D_A	uniform , unequal
Processing times	Distribution of the expected value across the sinks	D_{S_j}	uniform, unequal
Switching times	Distribution of the expected value across the sinks	D_{C_j}	uniform, unequal
	No switching time when switching from a sink to the queue with the same number	BV_2	boolean number

Table 7.4: Influencing parameters to select the appropriate service rule

The analyses show that the possible objectives pursued when selecting a service rule (see Table 7.3) can be summarized in groups. Within a group the same recommendations regarding the selection of the service rules apply. A distinction can be made between 5 clusters as shown in Table 7.5.

Time saving and performance of all queues	Time saving and performance of a specific queue k	Reduction of queueing of all queues	Reduction of a specific queue k	Balanced size of the queues
$\max(1/E(U))$	$\max(1/E(U_k))$	$\min(E(Q))$	$\min(E(\tilde{Q}_k))$	$\min(\max(E(\tilde{Q}_i)))$
$\min(E(U))$	$\min(E(U_k))$	$\min(Q_{0.95})$	$\min(\tilde{Q}_{k,0.95})$	$\min(\max(\tilde{Q}_{i,0.95}))$
$\min(U_{0.95})$	$\min(U_{k,0.95})$	$\min(p_{rejection})$	$\min(p_{k,rejection})$	
$\min(E(W))$	$\min(E(W_k))$			
$\min(W_{0.95})$	$\min(W_{k,0.95})$			

Table 7.5: Clustering of the possible objectives for selecting a service rule

Finally, recommendations can be given for the selection of suitable service rules. The following service rules have been identified as appropriate depending on the influencing parameters and the objectives pursued:

- Service rule 1: Select the next non-empty queue in order.
- Service rule 4: Select the queue with maximum length queue.
- Service rule 5: Select the queue with maximum waiting time.
- Service rule 6: Select the non-empty queue with minimum switching time.
- Service rule 7: Select the non-empty queue with minimum processing time.
- Service rule 9: Select the non-empty queue with the highest priority.
- Service rule 10: Select the queue according to absolutely prioritization.
- Service rule 14: Select the queue with the maximum queue length and serve the customers of this queue until it is empty.
- Service rule 15: Select the queue with maximum waiting time and serve the customers of this queue until it is empty.
- Service rule 17: Select the non-empty queue with minimum processing time and serve the customers of this queue until it is empty.

The recommendations for selecting one of these service rules based on the influencing parameters and the objectives pursued are presented in Table 7.6. The service rules which are not listed are not recommended at all.

Based on tables 7.4, 7.5 and 7.6, Table 7.7 is developed. Using this table, the appropriate service rules can be identified depending on the characteristics of a system and the objectives being pursued. In a subsequent step, these service rules can be modelled for the individual system using the analytical or simulation model of a MQSMDS from chapters 5 or 6 and thus analysed in detail. On the basis of the detailed analysis a final decision can be made regarding the service rule to be applied.

Service rule	Recommendation
Service rule 1	The use is only recommended if a certain sorting is to be achieved.
Service rule 4	When considering the reduction of queueing related to all queues or a balanced size of the queues as an objective, service rule 4 should be considered.
Service rule 5	When considering time saving related to all queues as an objective, service rule 5 should be considered.
Service rule 6	In case of unequal distribution of the expected value of the switching times across the queues, service rule 6 should be considered. For no switching time, when switching from a sink to the queue with the same number, service rule 6 should be considered.
Service rule 7	In case of unequal distribution of the expected value of the processing times across the queues, service rule 7 should be considered.
Service rule 9	When considering the performance parameters improvement of a specific queue as an objective, service rule 9 should be considered. When considering a balanced size of the queues as an objective and an unequal distribution of the expected values of the interarrival times over the queues, service rule 9 should be considered.
Service rule 10	When considering the performance parameters improvement of a specific queue as an objective, service rule 10 should be considered.
Service rule 14	For no switching time, when switching from a sink to the queue with the same number, deterministic transition probabilities across the sinks and uniform distribution of the expected values of the switching times across the queues, service rule 14 is to be considered instead of service rule 6. When considering a balanced size of the queues as an objective and an unequal distribution of the expected values of the interarrival times over the queues, service rule 14 should be considered.
Service rule 15	When considering time saving as an objective, no switching time, when switching from a sink to the queue with the same number, deterministic transition probabilities across the sinks and uniform distribution of the expected values of the switching times across the queues, service rule 15 should be considered.
Service rule 17	For no switching time, when switching from a sink to the queue with the same number, deterministic transition probabilities across the sinks, uniform distribution of the expected values of the switching times across the queues and unequal distribution of the expected value of the processing times across the queues, service rule 17 should be considered. When considering a balanced size of the queues as an objective, an unequal distribution of the expected values of the interarrival times over the queues and an unequal distribution of the expected values of the processing times across the queues, service rule 17 should be considered.

Table 7.6: Recommendation for selecting a service rule

D_P	D_A	D_S	D_C	BV_2	$\max(1/E(U))$ $\min(E(U))$ $\min(U_{0.95})$ $\min(E(W))$ $\min(W_{0.95})$	$\max(1/E(U_k))$ $\min(E(U_k))$ $\min(U_{k,0.95})$ $\min(E(W_k))$ $\min(W_{k,0.95})$	$\min(E(Q))$ $\min(Q_{0.95})$ $\min(p_{rejection})$	$\min(E(\tilde{Q}_k))$ $\min(\tilde{Q}_{k,0.95})$ $\min(p_{k,rejection})$	$\min(\max(E(\tilde{Q}_i)))$ $\min(\max(\tilde{Q}_{i,0.95}))$
uniform	uniform	uniform	uniform	no	5	9, 10	4	9, 10	4
uniform	uniform	uniform	uniform	yes	5, 6	6, 9, 10	4, 6	6, 9, 10	4, 6
uniform	uniform	uniform	unequal	no	5, 6	6, 9, 10	4, 6	6, 9, 10	4, 6
uniform	uniform	uniform	unequal	yes	5, 6	6, 9, 10	4, 6	6, 9, 10	4, 6
uniform	uniform	unequal	uniform	no	5, 7	7, 9, 10	4, 7	7, 9, 10	4, 7
uniform	uniform	unequal	uniform	yes	5, 6 , 7	6, 7, 9, 10	4, 6, 7	6, 7, 9, 10	4, 6, 7
uniform	uniform	unequal	unequal	no	5, 6 , 7	6, 7, 9, 10	4, 6, 7	6, 7, 9, 10	4, 6, 7
uniform	uniform	unequal	unequal	yes	5, 6 , 7	6, 7, 9, 10	4, 6, 7	6, 7, 9, 10	4, 6, 7
uniform	unequal	uniform	uniform	no	5	9, 10	4	9, 10	4, 9, 14
uniform	unequal	uniform	uniform	yes	5, 6	6, 9, 10	4, 6	6, 9, 10	4, 6, 9, 14
uniform	unequal	uniform	unequal	no	5, 6	6, 9, 10	4, 6	6, 9, 10	4, 6, 9, 14
uniform	unequal	uniform	unequal	yes	5, 6	6, 9, 10	4, 6	6, 9, 10	4, 6, 9, 14
uniform	unequal	unequal	uniform	no	5, 7	7, 9, 10	4, 7	7, 9, 10	4, 7, 9, 14, 17
uniform	unequal	unequal	uniform	yes	5, 6, 7	6, 7, 9, 10	4, 6, 7	6, 7, 9, 10	4, 6, 7, 9, 14, 17
uniform	unequal	unequal	unequal	no	5, 6, 7	6, 7, 9, 10	4, 6, 7	6, 7, 9, 10	4, 6, 7, 9, 14, 17
uniform	unequal	unequal	unequal	yes	5, 6, 7	6, 7, 9, 10	4, 6, 7	6, 7, 9, 10	4, 6, 7, 9, 14, 17

D_P	D_A	D_S	D_C	BV_2	$\max(1/E(U))$ $\min(E(U))$ $\min(U_{0.95})$ $\min(E(W))$ $\min(W_{0.95})$	$\max(1/E(U_k))$ $\min(E(U_k))$ $\min(U_{k,0.95})$ $\min(E(W_k))$ $\min(W_{k,0.95})$	$\min(E(Q))$ $\min(Q_{0.95})$ $\min(p_{rejection})$	$\min(E(\tilde{Q}_k))$ $\min(\tilde{Q}_{k,0.95})$ $\min(p_{k,rejection})$	$\min(\max(E(\tilde{Q}_i)))$ $\min(\max(\tilde{Q}_{i,0.95}))$
unequal	uniform	uniform	uniform	no	5	9, 10	4	9, 10	4
unequal	uniform	uniform	uniform	yes	5, 6	6, 9, 10	4, 6	6, 9, 10	4, 6
unequal	uniform	uniform	unequal	no	5, 6	6, 9, 10	4, 6	6, 9, 10	4, 6
unequal	uniform	uniform	unequal	yes	5, 6	6, 9, 10	4, 6	6, 9, 10	4, 6
unequal	uniform	unequal	uniform	no	5, 7	7, 9, 10	4, 7	7, 9, 10	4, 7
unequal	uniform	unequal	uniform	yes	5, 6, 7	6, 7, 9, 10	4, 6, 7	6, 7, 9, 10	4, 6, 7
unequal	uniform	unequal	unequal	no	5, 6, 7	6, 7, 9, 10	4, 6, 7	6, 7, 9, 10	4, 6, 7
unequal	uniform	unequal	unequal	yes	5, 6, 7	6, 7, 9, 10	4, 6, 7	6, 7, 9, 10	4, 6, 7
unequal	unequal	uniform	uniform	no	5	9, 10	4	9, 10	4, 9, 14
unequal	unequal	uniform	uniform	yes	5, 6	6, 9, 10	4, 6	6, 9, 10	4, 6, 9, 14
unequal	unequal	uniform	unequal	no	5, 6	6, 9, 10	4, 6	6, 9, 10	4, 6, 9, 14
unequal	unequal	uniform	unequal	yes	5, 6	6, 9, 10	4, 6	6, 9, 10	4, 6, 9, 14
unequal	unequal	unequal	uniform	no	5, 7	7, 9, 10	4, 7	7, 9, 10	4, 7, 9, 14, 17
unequal	unequal	unequal	uniform	yes	5, 6, 7	6, 7, 9, 10	4, 6, 7	6, 7,9, 10	4, 6, 7, 9, 14, 17
unequal	unequal	unequal	unequal	no	5, 6, 7	6, 7, 9, 10	4, 6, 7	6, 7,9, 10	4, 6, 7, 9, 14, 17
unequal	unequal	unequal	unequal	yes	5, 6, 7	6, 7, 9, 10	4, 6, 7	6, 7,9, 10	4, 6, 7, 9, 14, 17

D_P	D_A	D_S	D_C	BV_2	$\max(1/E(U))$ $\min(E(U))$ $\min(U_{0.95})$ $\min(E(W))$ $\min(W_{0.95})$	$\max(1/E(U_k))$ $\min(E(U_k))$ $\min(U_{k,0.95})$ $\min(E(W_k))$ $\min(W_{k,0.95})$	$\min(E(Q))$ $\min(Q_{0.95})$ $\min(p_{rejection})$	$\min(E(\tilde{Q}_k))$ $\min(\tilde{Q}_{k,0.95})$ $\min(p_{k,rejection})$	$\min(\max(E(\tilde{Q}_i)))$ $\min(\max(\tilde{Q}_{i,0.95}))$
determ.	uniform	uniform	uniform	no	5	9, 10	4	9, 10	4
determ.	uniform	uniform	uniform	yes	5, 14, 15	9, 10, 14, 15	4, 14	9, 10, 14	4, 14
determ.	uniform	uniform	unequal	no	5, 6	6, 9, 10	4, 6	6, 9, 10	4, 6
determ.	uniform	uniform	unequal	yes	5, 6	6, 9, 10	4, 6	6, 9, 10	4, 6
determ.	uniform	unequal	uniform	no	5, 7	7, 9, 10	4, 7	7, 9, 10	4, 7
determ.	uniform	unequal	uniform	yes	5, 7, 14, 15, 17	7, 9, 10, 14, 15, 17	4, 7, 14, 17	7, 9, 10, 14, 17	4, 7, 14, 17
determ.	uniform	unequal	unequal	no	5, 6, 7	6, 7, 9, 10	4, 6, 7	6, 7, 9, 10	4, 6, 7
determ.	uniform	unequal	unequal	yes	5, 6, 7	6, 7, 9, 10	4, 6, 7	6, 7, 9, 10	4, 6, 7
determ.	unequal	uniform	uniform	no	5	9, 10	4	9, 10	4, 9, 14
determ.	unequal	uniform	uniform	yes	5, 14, 15	9, 10, 14, 15	4, 14	9, 10, 14	4, 9, 14
determ.	unequal	uniform	unequal	no	5, 6	6, 9, 10	4, 6	6, 9, 10	4, 6, 9, 14
determ.	unequal	uniform	unequal	yes	5, 6	6, 9, 10	4, 6	6, 9, 10	4, 6, 9, 14
determ.	unequal	unequal	uniform	no	5, 7	7, 9, 10	4, 7	7, 9, 10	4, 7, 9, 14, 17
determ.	unequal	unequal	uniform	yes	5, 7, 14, 15, 17	7, 9, 10, 14, 15, 17	4, 7, 14, 17	7, 9, 10, 14, 17	4, 7, 9, 14, 17
determ.	unequal	unequal	unequal	no	5, 6, 7	6, 7, 9, 10	4, 6, 7	6, 7, 9, 10	4, 6, 7, 9, 14, 17
determ.	unequal	unequal	unequal	yes	5, 6, 7	6, 7, 9, 10	4, 6, 7	6, 7, 9, 10	4, 6, 7, 9, 14, 17

Table 7.7: Appropriate service rules depending on the influence parameters and the selected objective

7.3 Chapter Conclusion

In this chapter, the system characteristics of the MQSMDS are analysed, the service rules are evaluated and recommendations for the selection of a service rule are given.

The analysis of the system characteristics shows the special properties of the MQSMDS. Thus, with high interarrival time variabilities of the queues ($c^2(A_i) > 1$), the variability of the total stream is reduced by merging the steams when in contrast the total interarrival time variability is increased with low interarrival time variabilities of the queues ($c^2(A_i) < 1$). An unequal distribution of the expected value of the processing or switching times over the queues leads to higher variabilities in the system, which can be controlled by special service rules.

On the basis of the numerical evaluation of the service rules, recommendations can be made for the appropriate use of the service rules. Five influencing parameters out of 33 input parameters are identified, that affect the selection of the appropriate service rule. 18 possible objectives can be clustered in five groups, each with the same recommendation for the selection of a service rule. The number of relevant service rules are reduced to 10. Using Table 7.7 the appropriate service rules depending on the influence parameters and the selected objective can be determined. Thus, when choosing a service rule for a system, only a few different service rules have to be examined in detail and compared for the individual case.

8 Conclusion

Do to others what you
want them to do to you.
-Golden Rule, Bible

In this chapter the most important results of this research are summarized. In Section 8.1 a summary of this thesis is given on the basis of the research questions from Section 1.1. An outlook is presented in Section 8.2, in which further research areas are identified.

8.1 Summary

The objective of this work was to develop a modelling approach in discrete time domain in order to depict different service rules. The developed model is called *multi-queue system with multiple departure streams (MQSMDS)*. With this model the performance parameter distributions for different service rules can be calculated. The analysis and evaluations based on the model can be used to make recommendations about the appropriate use of service rules. With the results of this work a rapid and low-cost analysis and modelling of existing and planned specific material handling and production systems as well as a fast and easy identification of suitable service rules for these systems is possible.

The research of this work was divided into three parts with three research questions:

1. **Which service rules can be generally applied?**
2. **How can a discrete time queueing model with different service rules be modelled?**
3. **Which service rules should be used under which conditions?**

To answer research question 1, a holistic classification consisting of 2 rule categories, 7 rule classes and 16 rule types was created based on the literature from the various research and application areas. The combination of the different rule types results in a total of 480 service rules that can be modelled with this classification. Based on the resulting service rules of the classification a generic modelling of all important service rules of the various research and application areas is possible. With the presented assignment of the service rules of the different areas to the rule types of the classification, a service rule of an area can easily be modelled on the basis of the classification.

Research question 2 was answered by the development of a new discrete time queueing model, the *multi-queue system with multiple departure streams (MQSMDS)*. The model consists of N queues, one server and M sinks and can represent different service rules. The MQSMDS is modelled as a discrete time Markov chain. The basic system state can be defined as a $(2 \cdot N + 2)$-tuple consisting of the number of customers per queue, the residual interarrival time per queue, the queue of the next customer to be served and the sink of the last customer served. Depending on the selected service rule further random variables are added to the system state. The steady state distribution of the Markov chain is determined by solving the linear system of equations based on the transition probabilities.

The transition probabilities are calculated by summing the conditional probabilities from the following four parts:

- Calculation of the transition probability of queue lengths and residual interarrival time depending on a time interval.
- Calculation of the cycle time distribution.
- Modelling of service rules.
- Calculation of the transition probability of the sink of the last customer served.

The modelling of the service rules is based on the holistic classification. The performance parameters are determined by the transition probabilities and the steady state distribution. The distributions of the number of customers of the queues, waiting time of a customer of the queues, interdeparture time of the sinks, sojourn time from the queues to the sinks and total sojourn time are calculated. In addition, the utilization of the server and system as well as the rejection probabilities of the queues are determined.

Overall, a discrete time queueing model was developed

- with generally distributed interarrival and service times,
- which considers generally distributed switching times,
- whereby the time distributions can be different for each queue and sink,
- with which different service rule are modelled,
- with which all performance parameter distributions are determined for each queue,
- and with which the interdeparture time distribution is determined for different departure streams.

Furthermore, a simulation model was developed. It represents the MQSMDS using six events that can or must occur within a simulation iteration. It is necessary due to the limitation of the calculations with the analytical model. The validity is examined based on a comparison with the analytical model.

The third research question was answered with a numerical evaluation. The numerical study was performed using the discrete-event simulation model due to limited memory. To reflect a broad variety of different settings, the random variables whose distribution is specified as input in the MQSMDS are varied systematically. Altogether 630 different parameter settings are simulated and evaluated.

The analysis of the system characteristics shows the special properties of the MQSMDS. Thus, with high interarrival time variabilities of the queues ($c^2(A_i) > 1$) the variability of the total stream is reduced by merging the streams while the total interarrival time variability is increased with low interarrival time variabilities of the queues ($c^2(A_i) < 1$). An unequal distribution of the expected value of the processing or switching times over the queues leads to higher variabilities in the system, which can be controlled by special service rules.

On the basis of the numerical evaluation of the service rules, recommendations were made about the appropriate use of the service rules. Five influencing parameters out of 33 input parameters were identified, that affect the selection of the appropriate service rule. 18 possible objectives were clustered in five groups, each with the same recommendation for the selection of a service rule. The number of relevant service rules has been reduced from 480 to 10. The following service rules have been identified as appropriate depending on the characteristics of a system and the objectives being pursued:

- Service rule 1: Select the next non-empty queue in order.
- Service rule 4: Select the queue with maximum length queue.
- Service rule 5: Select the queue with maximum waiting time.
- Service rule 6: Select the non-empty queue with minimum switching time.
- Service rule 7: Select the non-empty queue with minimum processing time.

- Service rule 9: Select the non-empty queue with the highest priority.
- Service rule 10: Select the queue according to absolutely prioritization.
- Service rule 14: Select the queue with the maximum queue length and serve the customers of this queue until it is empty.
- Service rule 15: Select the queue with maximum waiting time and serve the customers of this queue until it is empty.
- Service rule 17: Select the non-empty queue with minimum processing time and serve the customers of this queue until it is empty.

The tables 7.6 and 7.7 summarize the use of these service rules depending on the influence parameters and the selected objective. Thus, when choosing a service rule for a system, only a few different service rules have to be examined in detail and compared for the individual case.

8.2 Outlook

With the presented discrete time queueing model different service rules can be depicted. The model is based on the assumption of a single server. However, this is not always the case depending on the area of application. In semiconductor manufacturing, for example, a production consists of machines with several chambers, which the products pass through in parallel. Furthermore, these are often machines with batch arrivals and batch processing. In order to be able to model such machines and to analyse the appropriate service rules, the model developed in this thesis must be extended accordingly. For the modelling of parallel machines the work of Matzka (2011) can be used. Batch arrivals and processing is analysed by Schleyer (2007). A combination of the model developed in this work with the models developed by Matzka (2011) and Schleyer (2007) should be addressed by further research.

Networks can be mapped using the decomposition approach. The interdeparture time distribution is used to connect two queueing systems, where the interdeparture time distribution of the first queueing system corresponds to the interarrival time distribution of the second system. For example, a production network with different products can be modelled using the developed model and the appropriate service rules can be analysed across the entire network. However, the decomposition approach assumes statistical independence of the operating systems in the network. This only applies to exponentially or geometrically distributed times. Otherwise a decomposition error occurs. In further research this error should be analysed and reduced to allow an accurate modelling of networks. First investigations concerning the decomposition error in discrete time domain were done by Jacobi (2018).

With the developed model, systems with a maximum of 50,000 states could be investigated with the available memory (64GB RAM). This limitation only allows the calculation of small systems, so that the simulation model had to be used for the numerical evaluation. Current research is constantly developing new general methods of main memory requirement and computational time reduction. For example, high bandwidth memory or hybrid memory cubes can significantly increase the data transfer rate of graphics memory (Windeck 2015). Together with the calculation of the models on graphics cards, the computing time could be significantly reduced. The application of such methods represents another point for further research.

Another possibility to use the developed models for larger systems is to reduce the number of random variables in the system state. For example, by omitting the residual interarrival time of the queues, the state space could almost be halved. Under the assumption of interarrival times independent of the last state of the system at the time of observation of the Markov chain, the steady state distribution and the performance parameters could still be determined.

However, this assumption applies only to exponentially/geometrically distributed interarrival times. For generally distributed interarrival times an approximate solution would result. This approximate approach is to be investigated and evaluated in terms of the approximation quality. Further heuristic approaches, which reduce the required memory and the computation time, for modelling different service rules based on the model developed in this thesis could be elaborated in further research.

Glossary of Notation

Basics

λ	Arrival rate of a queueing system	14
μ	Service rate of a queueing system	14
ρ	Utilization of a queueing system	14
σ_u	$u\%$-quantile of a random variable	11
$\vec{\pi}$	Stationary distribution of a Markov chain	13
$\vec{\pi}(n)$	State probability vector of a Markov chain at time n	13
$\vec{a}$	Interarrival time distribution	15
$\vec{b}$	Service time distribution	15
$\vec{x}$	Probability vector of a random variable X	10
A	Random variable of the interarrival time	15
a_n	Probability for an interarrival time of n	15
B	Random variable of the service time	15
b_m	Probability for an service time of m	15
c_X^2	Variance of a random variable X	11
$E(X)$	Expected value of a random variable X	11
I	State space of a Markov chain	12
i	Value of a discrete random variable	10

i_n	System state of a Markov chain at time n	12
i_{max}	Maximal value of a discrete random variable	10
L	Number of customers in a queueing system in a steady state	14
p_{ij}	Transition probability of a Markov chain from state i to state j	13
t_{inc}	Length of the discretization interval	10
$Var(X)$	Variance of a random variable X	11
W	Sojourn time in a queueing system in a steady state	14
X	Discrete random variable	10
X_n	Stochastic process of the Markov chain at time n	12
x_i	Probability that a discrete random variable X is equal to value i	10

Analytical Model

$\bar{e}_i^t$	Number of customers in queue i at time t without capacity restriction	98
$\ddot{\Delta}$	Time interval up to the time at which the condition to continue is no longer fulfilled	74
$\ddot{\Delta}_n$	Time interval up to the time at which the condition to continue is no longer fulfilled with rule type n of rule category 2	74
Δ	Cycle time	78
Δ^t	Time interval after the system state at time t	67
$\Delta^{n.1}$	Time interval between the start of time interval n and time $\tilde{t}$	89

$\Delta^{n.2}$	Time interval from time $\tilde{t}$ to the end of time interval Δ^n . . . 89
δ_j	Interdeparture time of of sink j . . . 93
Δ_{max}	Upper support of the cycle time . . . 91
η_i^t	Number of rejections of queue i at time t . . . 99
$\hat{p}_{i,j}$	Transition probability from queue i to sink j . . . 59
$\hat{P}_i$	Random variable of the transition from queue i to a sink . . . 59
$\hat{t}$	Time epoch . . . 59
ι_i	Number of arrivals in queue i in a cycle . . . 101
Λ	Random variable of the system state . . . 63
λ_x	Steady state probability for a system state x . . . 63
ω_i	Waiting time of a customer in queue i . . . 92
Π_n	Set of possible queues that have taken the largest/smallest/highest value . . . 83
Ψ	Auxiliary function for the additional rule for ambiguous decision . . . 83
ρ_{server}	Utilization of the server . . . 96
ρ_{system}	Utilization of the system . . . 97
τ	Point in time immediately before the start of a service . . . 63
Θ	Auxiliary function . . . 67
$\tilde{e}_i$	Number of customers in queue i at time $\tilde{t}$ . . . 90
$\tilde{P}$	Additional random variable for selecting a next queue . . . 82

$\tilde{p}_l$	Probability that queue l will be selected next based on $\vec{\tilde{p}}$	82
$\tilde{Q}_i$	Random variable of the number of customers in queue i at random epochs	88
$\tilde{q}_{i,n}$	Probability of a number of customers n in a queue i	91
$\tilde{t}$	Random point in time	89
ϑ	Total sojourn time	95
$\vec{\lambda}$	Steady state distribution	64
$\vec{\tilde{p}}$	Additional probability distribution for selecting a next queue	82
$\vec{\tilde{q}}_i$	Distribution of the number of customers in a queue i	91
$\vec{a}_i$	Interarrival time distribution of queue i	67
$\vec{c}_{ji}$	Switching time distribution from sink j to queue i	71
$\vec{d}_j^*$	Non-normalized interdeparture time distribution of sink j	94
$\vec{d}_j$	Interdeparture time distribution of sink j	94
$\vec{s}_{ij}$	Processing time distribution from queue i to sink j	72
$\vec{u}_{ij}$	Sojourn time distribution from a queue i to a sink j	94
$\vec{u}$	Total sojourn time distribution	95
$\vec{w}_i^*$	Non-normalized waiting time distribution of a customer in queue i	92
$\vec{w}_i$	Waiting time distribution of a customer in queue i	93
A_i	Random variable of the interarrival time of queue i	59
$a_{i,max}$	Upper support of the interarrival time of queue i	59
$a_{i,min}$	Lower support of the interarrival time of queue i	59

$a_{i,n}$	Interarrival time probability for value n of queue i	59
B	Random variable of the remaining time of the time window	63
b	Position in a queue	63
$c_{ji,max}$	Upper support of the switching time from sink j to queue i	59
$c_{ji,min}$	Lower support of the switching time from sink j to queue i	59
$c_{ji,n}$	Switching time probability for value n from sink j to queue i	59
C_{ji}	Random variable of the switching time from sink j to queue i	59
D_j	Random variable of the interdeparture time of sink j	88
d_{j,δ_j}	Probability of a interdeparture time of δ_j of sink j	94
d^*_{j,δ_j}	Non-normalized probability of a interdeparture time of δ_j of sink j	94
$d_{j,max}$	Upper support of the interdeparture time of sink j	63
$d_{j,min}$	Lower support of the interdeparture time of sink j	63
$E(A_i)$	Expected value of the interarrival time of queue i	95
$E(H_i)$	Expected value of the rejections of a queue i in a cycle	100
$E(L_i)$	Expected value of the arrivals in a queue i in a cycle	101
$E(T)$	Expected value of the cycle time	91
$E(T^t)$	Expected value of the time interval after the system state at time t	96

e_i	Number of customers in queue i at time τ 63
e_i^t	Number of customers in queue i at time t 67
f_i	Number of customers in queue i at time $\tau+1$ 63
G_i	Random variable of the sink of the first customer in queue i 63
g_i	Residual interarrival time of queue i at time τ 63
g_i^t	Residual interarrival time of queue i at time t 67
h_i	Residual interarrival time of queue i at time $\tau+1$ 63
i	Index of a queue 58
j	Index of a sink 58
k	Queue of the next customer to be served at time τ 63
K_i	Capacity of a queue i 58
l	Queue of the next customer to be served at time $\tau+1$ 63
L_j	Random variable of the last departure time of sink j 63
l_{max}	Maximum time between a last arrival and the end of a time interval depending on the waiting time of a customer 237
LV_i^1	Limit value with regard to the queue length of queue i 76
LV_i^2	Limit value with regard to the waiting time of queue i 77
M	Number of sinks of the MQSMDS 58
m_i	Sink of the first customer in queue i at time τ 63
m_{max}	Maximum time between a last arrival and the end of a time interval 67

$p_{continue}$	Probability of serving customers of the same queue again	78
p_{i,ω_i}	Probability of a waiting time of ω_i of a customer in queue i depending on time intervals Δ^1 and Δ^2	92
$p_{i,rejection}$	Rejection probability of a queue i	101
p_{j,δ_j}	Probability of a interdeparture time of δ_j of sink j depending on time intervals Δ^1, Δ^2 and Δ^3	93
$p_{n,continue}$	Probability of serving customers of the same queue again depending on rule type n of rule category 2	78
p_{next}	Probability that queue l will be selected next	78
$p_{RT_1,next}$	Probability that queue l will be selected next depending on rule type RT_1	81
p_{x^t}	Transition probability of queue lengths and residual interarrival times at time t	70
$p_{x^{t+1*}}$	Probability of transformation from $\bar{e}_i^t$ to e_i^t	99
p_{xy}	Transition probability from state x to state y	63
q	Remaining number of customers at time τ	63
Q_i	Random variable of the number of customers in queue i	62
r	Remaining number of customers at time $\tau + 1$	63
R_i	Random variable of the residual interarrival time of queue i	62
RT_1	Selected rule type of the rule category 1	73
RT_2	Set of selected rule types of the rule category 2	73
s_j	Last departure time of sink j at time τ	63
$s_{ij,max}$	Upper support of the processing time from queue i to sink j	59

$s_{ij,min}$	Lower support of the processing time from queue i to sink j 59
$s_{ij,n}$	Processing time probability for value n from queue i to sink j 59
S_{ij}	Random variable of the processing time from queue i to sink j 59
t	Point in time of an observation of a system state 67
t_j	Last departure time of sink j at time $\tau+1$ 63
TW_i	Fixed time window of queue i 85
TW_{max}	Maximum fixed time window over all queues 63
U	Total sojourn time 88
u	Sink of the last customer served at time τ 63
u_{ϑ}	Probability of a total sojourn time of ϑ 95
$u_{ij,max}$	Upper support of the sojourn time from queue i to sink j 88
$u_{ij,min}$	Lower support of the sojourn time from queue i to sink j 88
U_{ij}	Random variable of the sojourn time from queue i to sink j 88
u_{max}	Upper support of the total sojourn time 88
u_{min}	Lower support of the total sojourn time 88
v	Sink of the last customer served at time $\tau+1$ 63
W_i	Random variable of the waiting time of a customer in queue i 88
w_{i,ω_i}	Probability of a waiting time of ω_i of a customer in queue i 93

w^*_{i,ω_i}	Non-normalized probability of a waiting time of ω_i of a customer in queue i 92
$w_{i,max}$	Upper support of the waiting time in queue i.......... 63
W_i^b	Random variable of the waiting time of a customer at position b in queue i 63
x	System state.......... 63
x^t	System state at time t 70
x_{max}	State space size.......... 63
Y	Random variable of the queue of the next customer to be served.......... 62
y	System state.......... 63
y_i^b	Waiting time of a customer at position b in queue i at time τ 63
$y_i^{b^t}$	Waiting time of a customer at position b in queue i at time t.......... 69
Z	Random variable of the sink of the last customer served 62
z_i^b	Waiting time of a customer at position b in queue i at time $\tau+1$.......... 63

Validation

α	Probability of an error / significance level.......... 109
ε_1	Termination criterion of the simulation run.......... 108
ε_2	Termination criterion of the number of repetitions.......... 109
$\hat{p}_{i,j}$	Transition probability from queue i to sink j.......... 111
Λ^i	Sample of a simulation run from repetition i.......... 109

$\lvert\Delta^{abs}\rvert$	Absolute deviation	116
$\lvert\Delta^{rel}\rvert$	Relative deviation	116
$\tilde{\rho}$	Estimated utilization of the system	111
$\tilde{Q}^t$	Measurement of the total queue length at time t	108
d^*	Optimal truncation point	108
$E(\Lambda)(m)$	Estimator of the confidence interval for the expected value of the steady state distribution	109
$E(\tilde{A})$	Estimated expected value of total arrival time	111
$E(\tilde{C})$	Estimated expected value of total switching time	111
$E(\tilde{Q}(d,n))$	Expected values of the total queue length within the time interval from $(d+1)$ to n	108
$E(\tilde{S})$	Estimated expected value of total processing time	111
$E(A_i)$	Expected value of the interarrival time of queue i	111
$E(C_{ji})$	Expected value of the switching time from sink j to queue i	111
$E(S_{ij})$	Expected value of the processing time from queue i to sink j	111
e_i	Number of observations in class i	114
H_0	Null hypothesis of the chi-square test	114
I	State space size	103
i	Index of a classes	114
K	Number of classes	114
m	Number of repetitions of the simulation run	109
m_{min}	Minimum number of repetitions of the simulation run	109
$S^2(m)$	Empirical variance of the samples	109

T_1	Time interval of the start sample	108
T_2	Analysis interval during a simulation run	108
$t_{m-1,1-\alpha/2}$	Student's t-distribution	109

Evaluation

ρ_{server}	Utilization of the server	127
ρ_{system}	Utilization of the system	127
$\tilde{Q}_{0.95}$	95%-quantile of the total number of waiting customers	127
$\tilde{Q}_{i,0.95}$	95%-quantile of the number of waiting customers of queue i	127
$A_{0.95}$	95%-quantile of the total interarrival time	127
$A_{i,0.95}$	95%-quantile of the interarrival time of queue i	127
b_A	Gradient of the expected value of the interarrival times across the queues	122
$b_{\hat{P}}$	Gradient of the probability across the sinks	122
b_{C_i}	Gradient of the expected value of the switching times across the queues	122
b_{C_j}	Gradient of the expected value of the switching times across the sinks	122
b_{S_i}	Gradient of the expected value of the processing times across the queues	122
b_{S_j}	Gradient of the expected value of the processing times across the sinks	122
BV_1	Highest probability for the transition from a queue to the sink with the same number	122

BV_2	No switching time when switching from a sink to the queue with the same number	122
$c^2(\tilde{Q})$	Variability of the total number of waiting customers	127
$c^2(\tilde{Q}_i)$	Variability of the number of waiting customers of queue i	127
$c^2(A)$	Variability of the total interarrival time	127
$c^2(A_i)$	Variability of the interarrival times of queue i	122
$c^2(C)$	Variability of the total switching time	127
$c^2(C_i)$	Variability of the switching time of queue i	127
$c^2(C_{ji})$	Variability of the distribution of the switching times from sink i to queue j	122
$c^2(S)$	Variability of the total processing time	127
$c^2(S_i)$	Variability of the processing time of queue i	127
$c^2(S_{ij})$	Variability of the distribution of the processing times from queue $i$to sink j	122
$c^2(U)$	Variability of the total sojourn time	127
$c^2(U_i)$	Variability of the sojourn time of queue i	127
$c^2(W)$	Variability of the total waiting time of a customer	127
$c^2(W_i)$	Variability of the waiting time of a customer of queue i	127
$C_{0.95}$	95%-quantile of the total switching time	127
$C_{i,0.95}$	95%-quantile of the switching time of queue i	127
D_A	Distribution of the expected value of the interarrival times across the queues	122
$D_{\tilde{p}}$	Type of random distribution	122

D_{C_i}	Distribution of the expected value of the switching times across the queues 122
D_{C_j}	Distribution of the expected value of the switching times across the sinks 122
D_{S_i}	Distribution of the expected value of the processing times across the queues 122
D_{S_j}	Distribution of the expected value of the processing times across the sinks 122
$E(\tilde{A})$	Expected value of the total interarrival times 122
$E(\tilde{C})$	Expected value of the total switching times 122
$E(\tilde{Q})$	Expected value of the number of waiting customers of queue *i* 127
$E(\tilde{Q})$	Expected value of the total number of waiting customers 127
$E(\tilde{S})$	Expected value of the total processing times 122
$E(A)$	Expected value of the total interarrival time 127
$E(A_i)$	Expected value of the interarrival time of queue *i* 127
$E(C)$	Expected value of the total switching time 127
$E(C_i)$	Expected value of the switching time of queue *i* 127
$E(S)$	Expected value of the total processing time 127
$E(S_i)$	Expected value of the processing time of queue *i* 127
$E(U)$	Expected value of the total sojourn time 127
$E(U_i)$	Expected value of the sojourn time of queue *i* 127
$E(W)$	Expected value of the total waiting time of a customer 127

$E(W_i)$	Expected value of the waiting time of a customer of queue i	127
K_i	Queue capacities per queue	122
LV_i^1	Limit value of the queue length of queue$i$	122
LV_i^2	Limit value of the waiting time of queue$i$	122
M	Number of sinks	122
MN_i	Maximum number of services of queue$i$	122
N	Number of queues	122
$p_{i,rejection}$	Rejection probability of queue i	127
$p_{rejection}$	total rejection probability	127
$S_{0.95}$	95%-quantile of the total processing time	127
$S_{i,0.95}$	95%-quantile of the processing time of queue i	127
SR	Service rule number	122
T_A	Type of distribution of the total interarrival times	122
T_A	Type of distribution of the total processing times	122
T_A	Type of distribution of the total switching times	122
TW_i	Time window of queue$i$	122
$U_{0.95}$	95%-quantile of the total sojourn time	127
$U_{i,0.95}$	95%-quantile of the sojourn time of queue i	127
$W_{0.95}$	95%-quantile of the total waiting time of a customer	127
$W_{i,0.95}$	95%-quantile of the waiting time of a customer of queue i	127

Abbreviations

1LS	1-limited service	34
APRIO	Absolute priority	25
CDF	Cumulative distribution function	116
CL	Cyclic polling	21
CLS	Customer-limited service	22
DES	Discrete-event simulation	106
DP	Dynamic polling	21
DTMC	Discrete time Markov chain	12
EDD	Earliest due date	24
ES	Exhaustive service	22
FASFS	First arrived at shop, first served	24
FCFS	First come, first served	19
FG-FS	Fixed group with fixed sequence	25
FG-VS	Fixed group with variable sequence	25
FIFO	First-in-first-out	19
FRO	Fewest remaining operations	24
FTO	Fewest total operations	24
FTW-FS	Fixed time window with fixed sequence	25
FTW-VS	Fixed time window with variable sequence	25
G	Generally distributed times	27
G|G|1	Queueing system with generally distributed interarrival and service times and one single server	15
GDV	Greatest dollar value	24

GS	Gated service	22
HOL	Head of the line	19
LCFS	Last come, first served	19
LIP	Longest imminent processing time	24
LJF	Longest job first	19
LRP	Longest remaining processing time	24
LST	Laplace-Stieltjes transform	29
LTP	Longest total processing time	24
M	Exponentially distributed times	27
MAX	Maximum	110
MBPAP	Mean bounded priority with arrival pattern	31
MIN	Minimum	110
MQSMDS	Multi-queue system with multiple departure streams	58
MRO	Most remaining operations	24
MSER-5	Marginal standard error rule	107
MST	Minimum slack time	24
MTO	Most total operations	24
PP	Periodic polling	22
R&A	Research and application area	221
RC	Rule category	42
RP	Random polling	21
RPRIO	Relative priority	25
RS	Random service	19
RT	Rule type	42
SIP	Shortest imminent processing time	24

SJF	Shortest job first	19
SR	Service Rule	41
SRP	Shortest remaining processing time	24
STP	Shortest total processing time	24
StVO	Straßenverkehrsordnung / road traffic regulations	2
TLS	Time-limited service	22
VG-FS	Variable group with fixed sequence	25
VG-VS	Variable group with variable sequence	25
VTW-FS	Variable time window with fixed sequence	25
VTW-VS	Variable time window with variable sequence	25

References

Alfa, A. S. (1998). Discrete time analysis of MAP/PH/1 vacation queue with gated time-limited service. *Queueing Systems 29*(1), p. 35–54.

Alfa, A. S. (2003). Vacation models in discrete time. *Queueing Systems 44*(1), p. 5–30.

Alfa, A. S. (2010). *Queueing theory for telecommunications: discrete time modelling of a single node system.* Springer Science+Business Media, New York, NY, USA.

Arnold, D. and K. Furmans (2009). *Materialfluss in Logistiksystemen* (6. ed.). Springer, Berlin/Heidelberg, Germany.

Baker, J. and I. Rubin (1987). Polling with a general-service order table. *IEEE Transactions on Communications 35*(3), p. 283–288.

Beekhuizen, P. and J. A. C. Resing (2009). Approximation of discrete-time polling systems via structured markov chains. In: *Proceedings of the Fourth International ICST Conference on Performance Evaluation Methodologies and Tools*, p. 16–25.

Blackstone, J. H., D. T. Phillips and G. L. Hogg (1982). A state-of-the-art survey of dispatching rules for manufacturing job shop operations. *International Journal of Production Research 20*(1), p. 27–45.

Blanc, J. P. C. (1990). A numerical approach to cyclic-service queueing models. *Queueing Systems 6*(1), p. 173–188.

Bol, G. (2007). *Wahrscheinlichkeitstheorie : Einführung*. Lehrbuch. Oldenbourg Wissenschaftsverlag, München, Deutschland.

Bolch, G., S. Greiner, H. de Meer and K. S. Trivedi (2006). *Queueing networks and markov chains: modeling and performance evaluation with computer science applications*. John Wiley & Sons, Hoboken, NJ, USA.

Boon, M. A. and I. J. Adan (2009). Mixed gated/exhaustive service in a polling model with priorities. *Queueing Systems 63*(1-4), p. 383–399.

Boon, M. A. A., R. D. van der Mei and E. M. M. Winands (2011). Applications of polling systems. *Surveys in Operations Research and Management Science 16*(2), p. 67–82.

Boxma, O. J. (1991). Analysis and optimization of polling systems. In: C. D. e. Cohen, J. W. & Pack (eds.), *Queueing, Performance and Control in ATM (Proceedings 13th International Teletraffic Congress, ITC-13, June 19-26, 1991, Copenhagen, Denmark)*, p. 173–183.

Boxma, O. J. and W. P. Groenendijk (1988). Waiting times in discrete-time cyclic-service systems. *IEEE Transactions on Communications 36*(2), p. 164–170.

Boxma, O. J. and B. W. Meister (1987). Waiting-time approximations for cyclic-service systems with switchover times. *Performance Evaluation 7*(4), p. 299–308.

Browne, S. and U. Yechiali (1991). Dynamic scheduling in single-server multiclass service systems with unit buffers. *Naval Research Logistics (NRL) 38*(3), p. 383–396.

Bruneel, H. and B. G. Kim (1993). *Discrete-time models for communication systems including ATM*. Springer Science+Business Media, New York, NY, USA.

Chang, S. H. and D. W. Choi (2005). Performance analysis of a finite-buffer discrete-time queue with bulk arrival, bulk service and vacations. *Computers & operations research 32*(9), p. 2213–2234.

Conway, R. W., W. L. Maxwell and L. W. Miller (1967). *Theory of scheduling*. Dove Publications, Mineola, NY, USA.

Cooper, R. and G. Murray (1969). Queues served in cyclic order. *Bell System Technical Journal 48*(3), p. 675–689.

Cooper, R. B. (1970). Queues served in cyclic order: Waiting times. *Bell System Technical Journal 49*(3), p. 399–413.

Coury, S. and P. G. Harrison (1997). Waiting time distribution in a class of discrete-time cyclic service multi-queue systems. *Performance evaluation 29*(4), p. 223–243.

Cox, D. (2018). Watch your step: why the 10,000 daily goal is built on bad science. *The Guardian.* https://www.theguardian.com/lifeandstyle/2018/sep/03/watch-your-step-why-the-10000-daily-goal-is-built-on-bad-science (Accessed on 2018/12/21).

Derbala, A. (2005). Priority queuing in an operating system. *Computers & operations research 32*(2), p. 229–238.

Dittmann, R. and F. Hübner (1993). Discrete-time analysis of a cyclic service system with gated limited service. Technical report, University of Würzburg, Institute of Computer Science (Research Report Series), Würzburg, Germany.

Doshi, B. T. (1986). Queueing systems with vacations - a survey. *Queueing systems 1*(1), p. 29–66.

Engeln-Müllges, G., K. Niederdrenk and R. Wodicka (2011). *Numerik-Algorithmen: Verfahren, Beispiele, Anwendungen.* Springer, Berlin/Heidelberg, Germany.

Feller, W. (1970). *An introduction to probability theory and its applications* (3. ed.). New York: John Wiley & Sons, Hoboken, NJ, USA.

Frigui, I. and A. S. Alfa (1999). Analysis of a discrete time table polling system with MAP input and time-limited service discipline. *Telecommunication Systems 12*(1), p. 51–77.

Furmans, K. (2004). A framework of stochastic finite elements for models of material handling systems. In: *Progress in material handling research (Proceedings 8th International Material Handling Research Colloquium (IMHRC), June 13-17, 2004, Graz and Wels, Austria).*

Furmans, K., T. Schmidt, I. Meinhardt, M. Epp and M. Rimmele (2015). Analytical computation of sojourn time distributions in large-scale conveyer systems. In: *SMMSO 2015 (Proceedings 10th Conference on Stochastic Models of Manufacturing and Service Operations, June 1-6, 2015, Volos, Greece)*, p. 76–82.

Furmans, K. and A. Zillus (1996). Modeling independent production buffers in discrete time queueing networks. Technical report, University of Karlsruhe, IFK, Karlsruhe, Germany.

Grassmann, W. K. and J. L. Jain (1989). Numerical solutions of the waiting time distribution and idle time distribution of the arithmetic GI/G/1 queue. *Operations Research 37*(1), p. 141–150.

Großeschallau, W. (1984). *Materialflussrechnung: Modelle und Verfahren zur Analyse und Berechnung von Materialflusssystemen.* Springer, Berlin/Heidelberg, Germany.

Gudehus, T. (1976). Staueffekte vor Transportknoten. *Zeitschrift für Operations Research 20*(1), p. B207–B252.

Gudehus, T. (1977). *Transportsysteme für leichtes Stückgut*. VDI-Verlag, Düsseldorf, Germany.

Gudehus, T. and H. Kotzab (2012). *Comprehensive logistics*. Springer Science+Business Media, New York, NY, USA.

Günther, H.-O. and H. Tempelmeier (2005). *Produktion und Logistik*. Springer, Berlin/Heidelberg, Germany.

Hedtstück, U. (2013). *Simulation diskreter Prozesse*. Springer, Berlin/Heidelberg, Germany.

Heidemann, D. (1994). Queue length and delay distributions at traffic signals. *Transportation Research Part B: Methodological 28*(5), p. 377–389.

Heidemann, D. and H. Wegmann (1997). Queueing at unsignalized intersections. *Transportation Research Part B: Methodological 31*(3), p. 239–263.

Hoad, K. and S. Robinson (2011). Implementing MSER-5 in commercial simulation software and its wider implications. In: *WSC'11 (Proceedings of the Winter Simulation Conference, December 11-14, 2011, Phoenix, Arizona)*, p. 495–503.

Jacobi, C. (2018). *Analyse und Reduktion des Dekompositionsfehlers in zeidiskreten Bediensystemnetzwerken*. Master thesis supervised by M. Rimmele, Karlsruhe Institute of Technology, Institute for Material Handling and Logistics, Karlsruhe, Germany.

Jain, J. L. and W. K. Grassmann (1988). Numerical solution for the departure process from the GI/G/1 queue. *Computers & Operations Research 15*(3), p. 293–296.

Jaiswal, N. K. (1968). *Priority queues*. Academic Press, New York, USA.

Janker, H. (2018). *Straßenverkehrsrecht (StVR)*. dtv Verlagsgesellschaft, München, Germany.

Jolai, F., S. Asadzadeh and M. Taghizadeh (2010). A preemptive discrete-time priority buffer system with partial buffer sharing. *Applied Mathematical Modelling 34*(8), p. 2148–2165.

Kella, O. and U. Yechiali (1985). Waiting times in the non-preemptive priority M/M/c queue. *Stochastic Models 1*(2), p. 257–262.

Kella, O. and U. Yechiali (1988). Priorities in M/G/1 queue with server vacations. *Naval Research Logistics (NRL) 35*(1), p. 23–34.

Kendall, D. G. (1953). Stochastic processes occurring in the theory of queues and their analysis by the method of the imbedded markov chain. *The Annals of Mathematical Statistics 24*(3), p. 338–354.

Killat, U. (2011). *Entwurf und Analyse von Kommunikationsnetzen*, Volume 179. Springer, Berlin/Heidelberg, Germany.

Kleinrock, L. (1965). A conservation law for a wide class of queueing disciplines. *Naval Research Logistics Quarterly 12*(2), p. 181–192.

Kleinrock, L. (1975). *Queueing systems, Vol. 1: Theory*. John Wiley & Sons, Hoboken, NJ, USA.

Kleinrock, L. (1976). *Queueing systems, vol. 2: Computer applications*. John Wiley & Sons, Hoboken, NJ, USA.

Kleinrock, L. and H. Levy (1988). The analysis of random polling systems. *Operations Research 36*(5), p. 716–732.

Knott, G. D. (1975). Hashing functions. *The Computer Journal 18*(3), p. 265–278.

Korte, B. and J. Vygen (2012). *Kombinatorische Optimierung: Theorie und Algorithmen*. Springer, Berlin/Heidelberg, Germany.

Law, A. M., W. D. Kelton and W. D. Kelton (2015). *Simulation modeling and analysis*, Volume 2. McGraw-Hill ,New York, NY, USA.

Lee, Y. (2001). Discrete-time GeoX/G/1 queue with preemptive resume priority. *Mathematical and Computer Modelling 34*(3-4), p. 243–250.

Lee, Y., Y. H. Kim and J. D. Huh (2003). Discrete-time GeoX/G/1 que with non-preemptive priority. *Computers & Mathematics with Applications 46*(10-11), p. 1625–1632.

Leibowitz, M. A. (1968). Queues. *Scientific American 219*(2), p. 96–105.

Leung, K. K. (1990). Waiting time distributions for token-passing systems with limited-k services via discrete fourier transforms. In: *Performance '90 (Proceedings of the Tenth International Symposium on Computer Performance Modelling, Measurement and Evaluation, December 19-21, 1984, Paris, France)*, p. 333–347.

Leung, K. K. (1991). Cyclic-service systems with probabilistically-limited service. *IEEE Journal on Selected Areas in Communications 9*(2), p. 185–193.

Levy, H. and M. Sidi (1990). Polling systems: Applications, modeling, and optimization. *IEEE Transactions on Communications 38*(10), p. 1750–1760.

Little, J. D. C. (1961). A proof for the queuing Formula: L= lambda W. *Operations Research 9*(3), p. 383–387.

Lödding, H. (2005). *Verfahren der Fertigungssteuerung*, Volume 2. Springer, Berlin/Heidelberg, Germany.

Matzka, J. M. (2011). *Discrete time analysis of multi-server queueing systems in material handling and service.* PhD Thesis, Karlsruhe Institute of Technology, Institute for Material Handling and Logistics, Karlsruhe, Germany.

Nahmias, S. (2005). *Production and operations analysis* (5. ed.). Boston [u.a.]: McGraw-Hill ,New York, NY, USA.

Pinedo, M. (2009). *Planning and scheduling in manufacturing and services.* Springer, Berlin/Heidelberg, Germany.

Rimmele, M., K. Furmans and M. Epp (2015). Discrete-time analysis of a 4-way-crossing in conveying systems under the dispatching policy round robin. In: *SMMSO 2015 (Proceedings 10th Conference on Stochastic Models of Manufacturing and Service Operations, June 1-6, 2015, Volos, Greece)*, p. 195–202.

Schleyer, M. (2007). *Discrete time analysis of batch processes in material flow systems.* PhD Thesis, Karlsruhe Institute of Technology, Institute for Material Handling and Logistics, Karlsruhe, Germany.

Schleyer, M. and K. Furmans (2007). Performance evaluation of material flow systems in discrete time. *Logistics Journal 2007*((1)), p. 1–10.

Schwarz, J. A. and M. Epp (2016). Performance evaluation of a transportation-type bulk queue with generally distributed interarrival times. *International Journal of Production Research 54*(20), p. 6251–6264.

Shanthikumar, J. G. (1984). Analysis of priority queues with server control. *Opsearch 21*(3), p. 183–192.

Shanthikumar, J. G. (1989). Level crossing analysis of priority queues and a conservation identity for vacation models. *Naval Research Logistics (NRL) 36*(6), p. 797–806.

Shortle, J. F., J. M. Thompson, D. Gross and C. M. Harris (2018). *Fundamentals of queueing theory*. John Wiley & Sons, Hoboken, NJ, USA.

Takagi, H. (1986). *Analysis of polling systems*. MIT press, Cambridge, MA, USA.

Takagi, H. (1996). Bibliography on polling models. Technical report, Institute of Socio-Economic Planning, University of Tsukuba.

Takagi, H. (2000). Analysis and application of polling models. In: *Performance Evaluation: Origins and Directions*, p. 423–442. Springer, Berlin/Heidelberg, Germany.

Takagi, H. and L. Kleinrock (1984). A tutorial on the analysis of polling systems. Technical report, Computer Science Department, University of California, Los Angeles, California, USA.

Takagi, H. and K. K. Leung (1994). Analysis of a discrete-time queueing system with time-limited service. *Queueing Systems 18*(1-2), p. 183–197.

Takahashi, Y. and B. K. Kumar (1995). Pseudo-conservation law for discrete-time multi-queue systems with priority disciplines. *Journal of the Operations Research Society of Japan 38*(4), p. 450–466.

Tran-Gia, P. (1992). Analysis of polling systems with general input process and finite capacity. *IEEE Transactions on Communications 40*(2), p. 337–344.

Tran-Gia, P. (2005). *Einführung in die Leistungsbewertung und Verkehrstheorie*. Oldenbourg Wissenschaftsverlag, München, Deutschland.

Vollmann, T., W. Berry, D. C. Whybark and F. R. Jacobs (2004). *Manufacturing planning and control systems for supply chain management: the definitive guide for professionals*. McGraw-Hill ,New York, NY, USA.

Waldmann, K.-H. and U. M. Stocker (2013). *Stochastische Modelle: Eine anwendungsorientierte Einführung*. Springer, Berlin/Heidelberg, Germany.

Walraevens, J., B. Steyaert and H. Bruneel (2002). Delay characteristics in discrete-time GI-G-1 queues with non-preemptive priority queueing discipline. *Performance Evaluation 50*(1), p. 53–75.

Walraevens, J., B. Steyaert and H. Bruneel (2004). Performance analysis of a GI-Geo-1 buffer with a preemptive resume priority scheduling discipline. *European Journal of Operational Research 157*(1), p. 130–151.

Watson, K. S. (1985). Performance evaluation of cyclic service strategies - a survey. In: *Performance'84*, p. 521–533.

White, J. K. P. (1997). An effective truncation heuristic for bias reduction in simulation output. *Simulation 69*(6), p. 323–334.

Whitt, W. (1983). The queueing network analyzer. *The Bell System Technical Journal 62*(9), p. 2779–2815.

Windeck, C. (2015). Schnellstspeicher - HBM bringt Grafikkarten und Superrechner auf Trab. *c't 2015*(12), p. 140–143.

Wolff, R. W. (1989). *Stochastic modeling and the theory of queues*. Pearson Education, London, United Kingdom.

Zimmermann, O., M. Rimmele and K. Furmans (2018). The discrete-time G/G/1 queue with absolute priorities. *Working paper, Karlsruhe Institute of Technology, Institute for Material Handling and Logistics, Karlsruhe, Germany.*

List of Figures

List of Tables

A Service Rule Tables

In the following, the service rule tables are presented with regard to the classification of service rules presented in Chapter 4.

A.1 Table of the Service Rules Defined by the Rule Types from the Classification

Service rules	Service types															
	1.1	1.2	1.3	1.4	1.5	1.6	1.7	1.8	1.9	1.10	2.1	2.2	2.3	2.4	2.5	2.6
1	x															
2		x														
3			x													
4				x												
5					x											
6						x										
7							x									
8								x								
9									x							
10										x						
11	x										x					
12		x									x					
13			x								x					
14				x							x					
15					x						x					
16						x					x					
17							x				x					
18								x			x					
19									x		x					
20										x	x					
21	x											x				
22		x										x				
23			x									x				
24				x								x				
25					x							x				
26						x						x				
27							x					x				
28								x				x				
29									x			x				
30										x		x				
31	x												x			

Service rules	Service types															
	1.1	1.2	1.3	1.4	1.5	1.6	1.7	1.8	1.9	1.10	2.1	2.2	2.3	2.4	2.5	2.6
32		x											x			
33			x										x			
34				x									x			
35					x								x			
36						x							x			
37							x						x			
38								x					x			
39									x				x			
40										x			x			
41	x													x		
42		x												x		
43			x											x		
44				x										x		
45					x									x		
46						x								x		
47							x							x		
48								x						x		
49									x					x		
50										x				x		
51	x														x	
52		x													x	
53			x												x	
54				x											x	
55					x										x	
56						x									x	
57							x								x	
58								x							x	
59									x						x	
60										x					x	
61	x															x
62		x														x
63			x													x
64				x												x
65					x											x
66						x										x
67							x									x
68								x								x
69									x							x
70										x						x
71	x										x	x				
72		x									x	x				
73			x								x	x				
74				x							x	x				
75					x						x	x				
76						x					x	x				
77							x				x	x				
78								x			x	x				
79									x		x	x				
80										x	x	x				
81	x										x		x			
82		x									x		x			
83			x								x		x			
84				x							x		x			
85					x						x		x			
86						x					x		x			
87							x				x		x			
88								x			x		x			
89									x		x		x			
90										x	x		x			
91	x											x	x			
92		x										x	x			
93			x									x	x			

Service rules	Service types															
	1.1	1.2	1.3	1.4	1.5	1.6	1.7	1.8	1.9	1.10	2.1	2.2	2.3	2.4	2.5	2.6
94				x								x	x			
95					x							x	x			
96						x						x	x			
97							x					x	x			
98								x				x	x			
99									x			x	x			
100										x		x	x			
101	x											x		x		
102		x										x		x		
103			x									x		x		
104				x								x		x		
105					x							x		x		
106						x						x		x		
107							x					x		x		
108								x				x		x		
109									x			x		x		
110										x		x		x		
111	x												x	x		
112		x											x	x		
113			x										x	x		
114				x									x	x		
115					x								x	x		
116						x							x	x		
117							x						x	x		
118								x					x	x		
119									x				x	x		
120										x			x	x		
121	x										x				x	
122		x									x				x	
123			x								x				x	
124				x							x				x	
125					x						x				x	
126						x					x				x	
127							x				x				x	
128								x			x				x	
129									x		x				x	
130										x	x				x	
131	x											x			x	
132		x										x			x	
133			x									x			x	
134				x								x			x	
135					x							x			x	
136						x						x			x	
137							x					x			x	
138								x				x			x	
139									x			x			x	
140										x		x			x	
141	x												x		x	
142		x											x		x	
143			x										x		x	
144				x									x		x	
145					x								x		x	
146						x							x		x	
147							x						x		x	
148								x					x		x	
149									x				x		x	
150										x			x		x	
151	x													x	x	
152		x												x	x	
153			x											x	x	
154				x										x	x	
155					x									x	x	

Service rules	Service types															
	1.1	1.2	1.3	1.4	1.5	1.6	1.7	1.8	1.9	1.10	2.1	2.2	2.3	2.4	2.5	2.6
156						x								x	x	
157							x							x	x	
158								x						x	x	
159									x					x	x	
160										x				x	x	
161	x										x					x
162		x									x					x
163			x								x					x
164				x							x					x
165					x						x					x
166						x					x					x
167							x				x					x
168								x			x					x
169									x		x					x
170										x	x					x
171	x											x				x
172		x										x				x
173			x									x				x
174				x								x				x
175					x							x				x
176						x						x				x
177							x					x				x
178								x				x				x
179									x			x				x
180										x		x				x
181	x												x			x
182		x											x			x
183			x										x			x
184				x									x			x
185					x								x			x
186						x							x			x
187							x						x			x
188								x					x			x
189									x				x			x
190										x			x			x
191	x													x		x
192		x												x		x
193			x											x		x
194				x										x		x
195					x									x		x
196						x								x		x
197							x							x		x
198								x						x		x
199									x					x		x
200										x				x		x
201	x														x	x
202		x													x	x
203			x												x	x
204				x											x	x
205					x										x	x
206						x									x	x
207							x								x	x
208								x							x	x
209									x						x	x
210										x					x	x
211	x										x	x	x			
212		x									x	x	x			
213			x								x	x	x			
214				x							x	x	x			
215					x						x	x	x			
216						x					x	x	x			
217							x				x	x	x			

Service rules	Service types															
	1.1	1.2	1.3	1.4	1.5	1.6	1.7	1.8	1.9	1.10	2.1	2.2	2.3	2.4	2.5	2.6
218								x			x	x	x			
219									x		x	x	x			
220										x	x	x	x			
221	x											x	x	x		
222		x										x	x	x		
223			x									x	x	x		
224				x								x	x	x		
225					x							x	x	x		
226						x						x	x	x		
227							x					x	x	x		
228								x				x	x	x		
229									x			x	x	x		
230										x		x	x	x		
231	x										x	x			x	
232		x									x	x			x	
233			x								x	x			x	
234				x							x	x			x	
235					x						x	x			x	
236						x					x	x			x	
237							x				x	x			x	
238								x			x	x			x	
239									x		x	x			x	
240										x	x	x			x	
241	x										x		x		x	
242		x									x		x		x	
243			x								x		x		x	
244				x							x		x		x	
245					x						x		x		x	
246						x					x		x		x	
247							x				x		x		x	
248								x			x		x		x	
249									x		x		x		x	
250										x	x		x		x	
251	x											x	x		x	
252		x										x	x		x	
253			x									x	x		x	
254				x								x	x		x	
255					x							x	x		x	
256						x						x	x		x	
257							x					x	x		x	
258								x				x	x		x	
259									x			x	x		x	
260										x		x	x		x	
261	x											x		x	x	
262		x										x		x	x	
263			x									x		x	x	
264				x								x		x	x	
265					x							x		x	x	
266						x						x		x	x	
267							x					x		x	x	
268								x				x		x	x	
269									x			x		x	x	
270										x		x		x	x	
271	x												x	x	x	
272		x											x	x	x	
273			x										x	x	x	
274				x									x	x	x	
275					x								x	x	x	
276						x							x	x	x	
277							x						x	x	x	
278								x					x	x	x	
279									x				x	x	x	

Service rules	Service types															
	1.1	1.2	1.3	1.4	1.5	1.6	1.7	1.8	1.9	1.10	2.1	2.2	2.3	2.4	2.5	2.6
280										x			x	x	x	
281	x										x	x				x
282		x									x	x				x
283			x								x	x				x
284				x							x	x				x
285					x						x	x				x
286						x					x	x				x
287							x				x	x				x
288								x			x	x				x
289									x		x	x				x
290										x	x	x				x
291	x										x		x			x
292		x									x		x			x
293			x								x		x			x
294				x							x		x			x
295					x						x		x			x
296						x					x		x			x
297							x				x		x			x
298								x			x		x			x
299									x		x		x			x
300										x	x		x			x
301	x											x	x			x
302		x										x	x			x
303			x									x	x			x
304				x								x	x			x
305					x							x	x			x
306						x						x	x			x
307							x					x	x			x
308								x				x	x			x
309									x			x	x			x
310										x		x	x			x
311	x											x		x		x
312		x										x		x		x
313			x									x		x		x
314				x								x		x		x
315					x							x		x		x
316						x						x		x		x
317							x					x		x		x
318								x				x		x		x
319									x			x		x		x
320										x		x		x		x
321	x												x	x		x
322		x											x	x		x
323			x										x	x		x
324				x									x	x		x
325					x								x	x		x
326						x							x	x		x
327							x						x	x		x
328								x					x	x		x
329									x				x	x		x
330										x			x	x		x
331	x										x				x	x
332		x									x				x	x
333			x								x				x	x
334				x							x				x	x
335					x						x				x	x
336						x					x				x	x
337							x				x				x	x
338								x			x				x	x
339									x		x				x	x
340										x	x				x	x
341	x											x			x	x

Service rules	Service types															
	1.1	1.2	1.3	1.4	1.5	1.6	1.7	1.8	1.9	1.10	2.1	2.2	2.3	2.4	2.5	2.6
342		x										x			x	x
343			x									x			x	x
344				x								x			x	x
345					x							x			x	x
346						x						x			x	x
347							x					x			x	x
348								x				x			x	x
349									x			x			x	x
350										x		x			x	x
351	x												x		x	x
352		x											x		x	x
353			x										x		x	x
354				x									x		x	x
355					x								x		x	x
356						x							x		x	x
357							x						x		x	x
358								x					x		x	x
359									x				x		x	x
360										x			x		x	x
361	x													x	x	x
362		x												x	x	x
363			x											x	x	x
364				x										x	x	x
365					x									x	x	x
366						x								x	x	x
367							x							x	x	x
368								x						x	x	x
369									x					x	x	x
370										x				x	x	x
371	x										x	x	x		x	
372		x									x	x	x		x	
373			x								x	x	x		x	
374				x							x	x	x		x	
375					x						x	x	x		x	
376						x					x	x	x		x	
377							x				x	x	x		x	
378								x			x	x	x		x	
379									x		x	x	x		x	
380										x	x	x	x		x	
381	x											x	x	x	x	
382		x										x	x	x	x	
383			x									x	x	x	x	
384				x								x	x	x	x	
385					x							x	x	x	x	
386						x						x	x	x	x	
387							x					x	x	x	x	
388								x				x	x	x	x	
389									x			x	x	x	x	
390										x		x	x	x	x	
391	x										x	x	x			x
392		x									x	x	x			x
393			x								x	x	x			x
394				x							x	x	x			x
395					x						x	x	x			x
396						x					x	x	x			x
397							x				x	x	x			x
398								x			x	x	x			x
399									x		x	x	x			x
400										x	x	x	x			x
401	x											x	x	x		x
402		x										x	x	x		x
403			x									x	x	x		x

Service rules	Service types															
	1.1	1.2	1.3	1.4	1.5	1.6	1.7	1.8	1.9	1.10	2.1	2.2	2.3	2.4	2.5	2.6
404				x								x	x	x		x
405					x							x	x	x		x
406						x						x	x	x		x
407							x					x	x	x		x
408								x				x	x	x		x
409									x			x	x	x		x
410										x		x	x	x		x
411	x										x	x			x	x
412		x									x	x			x	x
413			x								x	x			x	x
414				x							x	x			x	x
415					x						x	x			x	x
416						x					x	x			x	x
417							x				x	x			x	x
418								x			x	x			x	x
419									x		x	x			x	x
420										x	x	x			x	x
421	x										x		x		x	x
422		x									x		x		x	x
423			x								x		x		x	x
424				x							x		x		x	x
425					x						x		x		x	x
426						x					x		x		x	x
427							x				x		x		x	x
428								x			x		x		x	x
429									x		x		x		x	x
430										x	x		x		x	x
431	x											x	x		x	x
432		x										x	x		x	x
433			x									x	x		x	x
434				x								x	x		x	x
435					x							x	x		x	x
436						x						x	x		x	x
437							x					x	x		x	x
438								x				x	x		x	x
439									x			x	x		x	x
440										x		x	x		x	x
441	x											x		x	x	x
442		x										x		x	x	x
443			x									x		x	x	x
444				x								x		x	x	x
445					x							x		x	x	x
446						x						x		x	x	x
447							x					x		x	x	x
448								x				x		x	x	x
449									x			x		x	x	x
450										x		x		x	x	x
451	x												x	x	x	x
452		x											x	x	x	x
453			x										x	x	x	x
454				x									x	x	x	x
455					x								x	x	x	x
456						x							x	x	x	x
457							x						x	x	x	x
458								x					x	x	x	x
459									x				x	x	x	x
460										x			x	x	x	x
461	x										x	x	x		x	x
462		x									x	x	x		x	x
463			x								x	x	x		x	x
464				x							x	x	x		x	x
465					x						x	x	x		x	x

Service rules	Service types															
	1.1	1.2	1.3	1.4	1.5	1.6	1.7	1.8	1.9	1.10	2.1	2.2	2.3	2.4	2.5	2.6
466						x					x	x	x		x	x
467							x				x	x	x		x	x
468								x			x	x	x		x	x
469									x		x	x	x		x	x
470										x	x	x	x		x	x
471	x											x	x	x	x	x
472		x										x	x	x	x	x
473			x									x	x	x	x	x
474				x								x	x	x	x	x
475					x							x	x	x	x	x
476						x						x	x	x	x	x
477							x					x	x	x	x	x
478								x				x	x	x	x	x
479									x			x	x	x	x	x
480										x		x	x	x	x	x

Table A.1: Service rules defined by the rule types from the classification

A.2 Table of the Service Rules from the Various Research and Application Areas Assigned to the Defined Service Rules

R&A rules	Service rules	Service types															
		1.1	1.2	1.3	1.4	1.5	1.6	1.7	1.8	1.9	1.10	2.1	2.2	2.3	2.4	2.5	2.6
FCFS	5					x											
LCFS	9									x							
RS	3			x													
SJF	7							x									
LJF	8								x								
HOL	9									x							
CP-ES	12		x									x					
CP-GS	42		x												x		
CP-TLS	22		x										x				
CP-CLS	32		x											x			
RP-ES	13			x								x					
RP-GS	43			x											x		
RP-TLS	23			x									x				
RP-CLS	33			x										x			
DP-ES	14				x							x					
	15					x						x					
	16						x					x					
	17							x				x					
	18								x			x					
	19									x		x					
DP-GS	44				x										x		
	45					x									x		
	46						x								x		
	47							x							x		
	48								x						x		
	49									x					x		
DP-TLS	24				x								x				
	25					x							x				
	26						x						x				
	27							x					x				
	28								x				x				
	29									x			x				

R&A rules	Service rules	Service types															
		1.1	1.2	1.3	1.4	1.5	1.6	1.7	1.8	1.9	1.10	2.1	2.2	2.3	2.4	2.5	2.6
DP-CLS	34				x									x			
	35					x								x			
	36						x							x			
	37							x						x			
	38								x					x			
	39									x				x			
FRO	9									x							
MRO	9									x							
FTO	9									x							
MTO	9									x							
SIP	9									x							
LIP	7							x									
SRP	8								x								
LRP	9									x							
STP	9									x							
LTP	9									x							
EDD	9									x							
MST	9									x							
RS	9									x							
FCFS	5					x											
FASFS	9									x							
GDV	9									x							
FCFS	5					x											
RS	3			x													
FTW-FS	21	x											x				
FTW-VS	22		x										x				
	23			x									x				
	24				x								x				
	25					x							x				
	26						x						x				
	27							x					x				
	28								x				x				
	29									x			x				
	30										x		x				
VTW-FS	71	x										x	x				
	91	x											x	x			
	101	x											x		x		
	131	x											x			x	
	171	x											x				x
	211	x										x	x	x			
	221	x											x	x	x		
	231	x										x	x			x	
	251	x											x	x		x	
	261	x											x		x	x	
	281	x										x	x				x
	301	x											x	x			x
	311	x											x		x		x
	341	x											x			x	x
	371	x										x	x	x		x	
	381	x											x	x	x	x	
	391	x										x	x	x			x
	401	x											x	x	x		x
	411	x										x	x			x	x
	431	x											x	x		x	x
	441	x											x		x	x	x
	461	x										x	x	x		x	x
	471	x											x	x	x	x	x
VTW-VS	22		x										x				
	23			x									x				
	24				x								x				
	25					x							x				
	26						x						x				
	27							x					x				
	28								x				x				

R&A rules	Service rules	Service types															
		1.1	1.2	1.3	1.4	1.5	1.6	1.7	1.8	1.9	1.10	2.1	2.2	2.3	2.4	2.5	2.6
	29									x			x				
	30										x		x				
	72		x									x	x				
	73			x								x	x				
	74				x							x	x				
	75					x						x	x				
	76						x					x	x				
	77							x				x	x				
	78								x			x	x				
	79									x		x	x				
	80										x	x	x				
	92		x										x	x			
	93			x									x	x			
	94				x								x	x			
	95					x							x	x			
	96						x						x	x			
	97							x					x	x			
	98								x				x	x			
	99									x			x	x			
	100										x		x	x			
	102		x										x		x		
	103			x									x		x		
	104				x								x		x		
	105					x							x		x		
	106						x						x		x		
	107							x					x		x		
	108								x				x		x		
	109									x			x		x		
	110										x		x		x		
	132		x										x			x	
	133			x									x			x	
	134				x								x			x	
	135					x							x			x	
	136						x						x			x	
	137							x					x			x	
	138								x				x			x	
	139									x			x			x	
	140										x		x			x	
	172		x										x				x
	173			x									x				x
	174				x								x				x
	175					x							x				x
	176						x						x				x
	177							x					x				x
	178								x				x				x
	179									x			x				x
	180										x		x				x
	212		x									x	x	x			
	213			x								x	x	x			
	214				x							x	x	x			
	215					x						x	x	x			
	216						x					x	x	x			
	217							x				x	x	x			
	218								x			x	x	x			
	219									x		x	x	x			
	220										x	x	x	x			
	222		x										x	x	x		
	223			x									x	x	x		
	224				x								x	x	x		
	225					x							x	x	x		
	226						x						x	x	x		
	227							x					x	x	x		
	228								x				x	x	x		
	229									x			x	x	x		
	230										x		x	x	x		

R&A rules	Service rules	Service types															
		1.1	1.2	1.3	1.4	1.5	1.6	1.7	1.8	1.9	1.10	2.1	2.2	2.3	2.4	2.5	2.6
	232		x									x	x			x	
	233			x								x	x			x	
	234				x							x	x			x	
	235					x						x	x			x	
	236						x					x	x			x	
	237							x				x	x			x	
	238								x			x	x			x	
	239									x		x	x			x	
	240										x	x	x			x	
	252		x										x	x		x	
	253			x									x	x		x	
	254				x								x	x		x	
	255					x							x	x		x	
	256						x						x	x		x	
	257							x					x	x		x	
	258								x				x	x		x	
	259									x			x	x		x	
	260										x		x	x		x	
	262		x										x		x	x	
	263			x									x		x	x	
	264				x								x		x	x	
	265					x							x		x	x	
	266						x						x		x	x	
	267							x					x		x	x	
	268								x				x		x	x	
	269									x			x		x	x	
	270										x		x		x	x	
	282		x									x	x				x
	283			x								x	x				x
	284				x							x	x				x
	285					x						x	x				x
	286						x					x	x				x
	287							x				x	x				x
	288								x			x	x				x
	289									x		x	x				x
	290										x	x	x				x
	302		x										x	x			x
	303			x									x	x			x
	304				x								x	x			x
	305					x							x	x			x
	306						x						x	x			x
	307							x					x	x			x
	308								x				x	x			x
	309									x			x	x			x
	310										x		x	x			x
	312		x										x		x		x
	313			x									x		x		x
	314				x								x		x		x
	315					x							x		x		x
	316						x						x		x		x
	317							x					x		x		x
	318								x				x		x		x
	319									x			x		x		x
	320										x		x		x		x
	342		x										x			x	x
	343			x									x			x	x
	344				x								x			x	x
	345					x							x			x	x
	346						x						x			x	x
	347							x					x			x	x
	348								x				x			x	x
	349									x			x			x	x
	350										x		x			x	x
	372		x									x	x	x		x	
	373			x								x	x	x		x	

R&A rules	Service rules	Service types															
		1.1	1.2	1.3	1.4	1.5	1.6	1.7	1.8	1.9	1.10	2.1	2.2	2.3	2.4	2.5	2.6
	374				x							x	x	x		x	
	375					x						x	x	x		x	
	376						x					x	x	x		x	
	377							x				x	x	x		x	
	378								x			x	x	x		x	
	379									x		x	x	x		x	
	380										x	x	x	x		x	
	382		x										x	x	x	x	
	383			x									x	x	x	x	
	384				x								x	x	x	x	
	385					x							x	x	x	x	
	386						x						x	x	x	x	
	387							x					x	x	x	x	
	388								x				x	x	x	x	
	389									x			x	x	x	x	
	390										x		x	x	x	x	
	392		x									x	x	x			x
	393			x								x	x	x			x
	394				x							x	x	x			x
	395					x						x	x	x			x
	396						x					x	x	x			x
	397							x				x	x	x			x
	398								x			x	x	x			x
	399									x		x	x	x			x
	400										x	x	x	x			x
	402		x										x	x	x		x
	403			x									x	x	x		x
	404				x								x	x	x		x
	405					x							x	x	x		x
	406						x						x	x	x		x
	407							x					x	x	x		x
	408								x				x	x	x		x
	409									x			x	x	x		x
	410										x		x	x	x		x
	412		x									x	x			x	x
	413			x								x	x			x	x
	414				x							x	x			x	x
	415					x						x	x			x	x
	416						x					x	x			x	x
	417							x				x	x			x	x
	418								x			x	x			x	x
	419									x		x	x			x	x
	420										x	x	x			x	x
	432		x										x	x		x	x
	433			x									x	x		x	x
	434				x								x	x		x	x
	435					x							x	x		x	x
	436						x						x	x		x	x
	437							x					x	x		x	x
	438								x				x	x		x	x
	439									x			x	x		x	x
	440										x		x	x		x	x
	442		x										x		x	x	x
	443			x									x		x	x	x
	444				x								x		x	x	x
	445					x							x		x	x	x
	446						x						x		x	x	x
	447							x					x		x	x	x
	448								x				x		x	x	x
	449									x			x		x	x	x
	450										x		x		x	x	x
	462		x									x	x	x		x	x
	463			x								x	x	x		x	x
	464				x							x	x	x		x	x
	465					x						x	x	x		x	x

R&A rules	Service rules	Service types															
		1.1	1.2	1.3	1.4	1.5	1.6	1.7	1.8	1.9	1.10	2.1	2.2	2.3	2.4	2.5	2.6
	466						x					x	x	x		x	x
	467							x				x	x	x		x	x
	468								x			x	x	x		x	x
	469									x		x	x	x		x	x
	470										x	x	x	x		x	x
	472		x										x	x	x	x	x
	473			x									x	x	x	x	x
	474				x								x	x	x	x	x
	475					x							x	x	x	x	x
	476						x						x	x	x	x	x
	477							x					x	x	x	x	x
	478								x				x	x	x	x	x
	479									x			x	x	x	x	x
	480										x		x	x	x	x	x
FG-FS	31	x												x			
FG-VS	32		x											x			
	33			x										x			
	34				x									x			
	35					x								x			
	36						x							x			
	36						x							x			
	38								x					x			
	39									x				x			
	40										x			x			
VG-FS	81	x										x		x			
	91	x											x	x			
	111	x												x	x		
	141	x												x		x	
	181	x												x			x
	211	x										x	x	x			
	221	x											x	x	x		
	241	x										x		x		x	
	251	x											x	x		x	
	271	x												x	x	x	
	291	x										x		x			x
	301	x											x	x			x
	321	x												x	x		x
	351	x												x		x	x
	371	x										x	x	x		x	
	381	x											x	x	x	x	
	391	x										x	x	x			x
	401	x											x	x	x		x
	421	x										x		x		x	x
	431	x											x	x		x	x
	451	x												x	x	x	x
	461	x										x	x	x		x	x
	471	x											x	x	x	x	x
VG-VS	82		x									x		x			
	83			x								x		x			
	84				x							x		x			
	85					x						x		x			
	86						x					x		x			
	87							x				x		x			
	88								x			x		x			
	89									x		x		x			
	90										x	x		x			
	92		x										x	x			
	93			x									x	x			
	94				x								x	x			
	95					x							x	x			
	96						x						x	x			
	97							x					x	x			
	98								x				x	x			
	99									x			x	x			
	100										x		x	x			

R&A rules	Service rules	Service types															
		1.1	1.2	1.3	1.4	1.5	1.6	1.7	1.8	1.9	1.10	2.1	2.2	2.3	2.4	2.5	2.6
	112		x											x	x		
	113			x										x	x		
	114				x									x	x		
	115					x								x	x		
	116						x							x	x		
	117							x						x	x		
	118								x					x	x		
	119									x				x	x		
	120										x			x	x		
	142		x											x		x	
	143			x										x		x	
	144				x									x		x	
	145					x								x		x	
	146						x							x		x	
	147							x						x		x	
	148								x					x		x	
	149									x				x		x	
	150										x			x		x	
	182		x											x			x
	183			x										x			x
	184				x									x			x
	185					x								x			x
	186						x							x			x
	187							x						x			x
	188								x					x			x
	189									x				x			x
	190										x			x			x
	212		x									x	x	x			
	213			x								x	x	x			
	214				x							x	x	x			
	215					x						x	x	x			
	216						x					x	x	x			
	217							x				x	x	x			
	218								x			x	x	x			
	219									x		x	x	x			
	220										x	x	x	x			
	222		x										x	x	x		
	223			x									x	x	x		
	224				x								x	x	x		
	225					x							x	x	x		
	226						x						x	x	x		
	227							x					x	x	x		
	228								x				x	x	x		
	229									x			x	x	x		
	230										x		x	x	x		
	242		x									x		x		x	
	243			x								x		x		x	
	244				x							x		x		x	
	245					x						x		x		x	
	246						x					x		x		x	
	247							x				x		x		x	
	248								x			x		x		x	
	249									x		x		x		x	
	250										x	x		x		x	
	252		x										x	x		x	
	253			x									x	x		x	
	254				x								x	x		x	
	255					x							x	x		x	
	256						x						x	x		x	
	257							x					x	x		x	
	258								x				x	x		x	
	259									x			x	x		x	
	260										x		x	x		x	
	272		x											x	x	x	
	273			x										x	x	x	

R&A rules	Service rules	Service types															
		1.1	1.2	1.3	1.4	1.5	1.6	1.7	1.8	1.9	1.10	2.1	2.2	2.3	2.4	2.5	2.6
	274				x									x	x	x	
	275					x								x	x	x	
	276						x							x	x	x	
	277							x						x	x	x	
	278								x					x	x	x	
	279									x				x	x	x	
	280										x			x	x	x	
	292		x									x		x			x
	293			x								x		x			x
	294				x							x		x			x
	295					x						x		x			x
	296						x					x		x			x
	297							x				x		x			x
	298								x			x		x			x
	299									x		x		x			x
	300										x	x		x			x
	302		x										x	x			x
	303			x									x	x			x
	304				x								x	x			x
	305					x							x	x			x
	306						x						x	x			x
	307							x					x	x			x
	308								x				x	x			x
	309									x			x	x			x
	310										x		x	x			x
	322		x											x	x		x
	323			x										x	x		x
	324				x									x	x		x
	325					x								x	x		x
	326						x							x	x		x
	327							x						x	x		x
	328								x					x	x		x
	329									x				x	x		x
	330										x			x	x		x
	352		x											x		x	x
	353			x										x		x	x
	354				x									x		x	x
	355					x								x		x	x
	356						x							x		x	x
	357							x						x		x	x
	358								x					x		x	x
	359									x				x		x	x
	360										x			x		x	x
	372		x									x	x	x		x	
	373			x								x	x	x		x	
	374				x							x	x	x		x	
	375					x						x	x	x		x	
	376						x					x	x	x		x	
	377							x				x	x	x		x	
	378								x			x	x	x		x	
	379									x		x	x	x		x	
	380										x	x	x	x		x	
	382		x										x	x	x	x	
	383			x									x	x	x	x	
	384				x								x	x	x	x	
	385					x							x	x	x	x	
	386						x						x	x	x	x	
	387							x					x	x	x	x	
	388								x				x	x	x	x	
	389									x			x	x	x	x	
	390										x		x	x	x	x	
	392		x									x	x	x			x
	393			x								x	x	x			x
	394				x							x	x	x			x
	395					x						x	x	x			x

R&A rules	Service rules	Service types															
		1.1	1.2	1.3	1.4	1.5	1.6	1.7	1.8	1.9	1.10	2.1	2.2	2.3	2.4	2.5	2.6
	396						x					x	x	x			x
	397							x				x	x	x			x
	398								x			x	x	x			x
	399									x		x	x	x			x
	400										x	x	x	x			x
	402		x										x	x	x		x
	403			x									x	x	x		x
	404				x								x	x	x		x
	405					x							x	x	x		x
	406						x						x	x	x		x
	407							x					x	x	x		x
	408								x				x	x	x		x
	409									x			x	x	x		x
	410										x		x	x	x		x
	422		x									x		x		x	x
	423			x								x		x		x	x
	424				x							x		x		x	x
	425					x						x		x		x	x
	426						x					x		x		x	x
	427							x				x		x		x	x
	428								x			x		x		x	x
	429									x		x		x		x	x
	430										x	x		x		x	x
	432		x										x	x		x	x
	433			x									x	x		x	x
	434				x								x	x		x	x
	435					x							x	x		x	x
	436						x						x	x		x	x
	437							x					x	x		x	x
	438								x				x	x		x	x
	439									x			x	x		x	x
	440										x		x	x		x	x
	452		x											x	x	x	x
	453			x										x	x	x	x
	454				x									x	x	x	x
	455					x								x	x	x	x
	456						x							x	x	x	x
	457							x						x	x	x	x
	458								x					x	x	x	x
	459									x				x	x	x	x
	460										x			x	x	x	x
	462		x									x	x	x		x	x
	463			x								x	x	x		x	x
	464				x							x	x	x		x	x
	465					x						x	x	x		x	x
	466						x					x	x	x		x	x
	467							x				x	x	x		x	x
	468								x			x	x	x		x	x
	469									x		x	x	x		x	x
	470										x	x	x	x		x	x
	472		x										x	x	x	x	x
	473			x									x	x	x	x	x
	474				x								x	x	x	x	x
	475					x							x	x	x	x	x
	476						x						x	x	x	x	x
	477							x					x	x	x	x	x
	478								x				x	x	x	x	x
	479									x			x	x	x	x	x
	480										x		x	x	x	x	x
RPRIO	9									x							
APRIO	10										x						

Table A.2: Assignment of the service rules from various research and application areas to the defined service rules from the classification

B Additional Mathematical Definitions and Calculations

In the following, additional mathematical definitions and calculations are given with regard to the analytical model presented in Chapter 5 and the numerical evaluation in Chapter 7.

B.1 Definitions of the Extended System States

- System state extended by the waiting time of a customer W_i^b at position $b = \{1, 2, \cdots, K_i\}$ in the queue $i = \{1, 2, \cdots, N\}$:

$$
\begin{aligned}
&(Q_1, \ldots, Q_N, R_1, \ldots, R_N, Y, Z, W_1^1, \ldots, W_N^1, \ldots, W_1^{K_1}, \ldots W_N^{K_N}) \\
&\quad \text{with} \\
&\quad Q_i \in \{1, ..., K_i\} \quad i \in \{1, \ldots, N\} \\
&\quad R_i \in \left\{1, ..., a_{i,max}\right\} \quad i \in \{1, \ldots, N\} \\
&\quad Y \in \{1, ..., N\} \\
&\quad Z \in \{1, ..., M\} \\
&\quad W_i^b \in \left\{1, \ldots, w_{i,max}^b\right\} \quad i \in \{1, \ldots, N\} \quad b \in \{1, \ldots, K_i\}
\end{aligned}
\tag{B.1}
$$

- System state extended by the remaining time of the time window B:

$$
\begin{aligned}
&(Q_1,\ldots,Q_N,R_1,\ldots,R_N,Y,Z,B) \\
&\quad \text{with} \\
&\quad Q_i \in \{1,...,K_i\} \quad i \in \{1,\ldots,N\} \\
&\quad R_i \in \left\{1,...,a_{i,max}\right\} \quad i \in \{1,\ldots,N\} \\
&\quad Y \in \{1,...,N\} \\
&\quad Z \in \{1,...,M\} \\
&\quad B \in \{1,...,TW_{max}\}
\end{aligned}
\tag{B.2}
$$

- System state extended by the remaining number of customers G:

$$
\begin{aligned}
&(Q_1,\ldots,Q_N,R_1,\ldots,R_N,Y,Z,G) \\
&\quad \text{with} \\
&\quad Q_i \in \{1,...,K_i\} \quad i \in \{1,\ldots,N\} \\
&\quad R_i \in \left\{1,...,a_{i,max}\right\} \quad i \in \{1,\ldots,N\} \\
&\quad Y \in \{1,...,N\} \\
&\quad Z \in \{1,...,M\} \\
&\quad G \in \{1,...,MN_{max}\}
\end{aligned}
\tag{B.3}
$$

- System state extended by the sink of the first customer O_i in the queue $i = \{1,2,\cdots,N\}$:

$$
\begin{aligned}
&(Q_1,\ldots,Q_N,R_1,\ldots,R_N,Y,Z,O_1,\ldots,O_N) \\
&\quad \text{with} \\
&\quad Q_i \in \{1,...,K_i\} \quad i \in \{1,\ldots,N\} \\
&\quad R_i \in \left\{1,...,a_{i,max}\right\} \quad i \in \{1,\ldots,N\} \\
&\quad Y \in \{1,...,N\} \\
&\quad Z \in \{1,...,M\} \\
&\quad O_i \in \{1,...,M\} \quad i \in \{1,\ldots,N\}
\end{aligned}
\tag{B.4}
$$

- System state extended by the last departure time L_j per sink $j \in \{1,...,M\}$:

$$
\begin{aligned}
&(Q_1,\dots,Q_N,R_1,\dots,R_N,Y,Z,L_1,\dots,L_M)\\
&\quad\text{with}\\
&\quad Q_i \in \{1,...,K_i\} \quad i \in \{1,\dots,N\}\\
&\quad R_i \in \{1,...,a_{i,max}\} \quad i \in \{1,\dots,N\}\\
&\quad Y \in \{1,...,N\}\\
&\quad Z \in \{1,...,M\}\\
&\quad L_j \in \{1,...,d_{j,max}\} \quad j \in \{1,\dots,M\}
\end{aligned}
\tag{B.5}
$$

B.2 Definition of the Transition Probability for the Extended System States

- Transition probability for a system state extended by the waiting time of a customer W_i^b at position $b = \{1,2,\cdots,K_i\}$ in the queue $i = \{1,2,\cdots,N\}$:

$$
\begin{aligned}
p_{xy} =& p_{(e_1,\dots,e_n,g_1,\dots,g_n,k,u,y_1^1,\dots,y_N^1,\dots,y_1^{K_1},\dots y_N^{K_N}),(f_1,\dots,f_n,h_1,\dots,h_n,l,v,z_1^1,\dots,z_N^1,\dots,z_1^{K_1},\dots z_N^{K_N})}\\
=& P(\Lambda^{\tau+1} = y \mid \Lambda^{\tau} = x)\\
=& P(Q_1^{\tau+1} = f_1,\dots,Q_n^{\tau+1} = f_n, R_1^{\tau+1} = h_1,\dots,R_n^{\tau+1} = h_n, Y^{\tau+1} = l,\\
& Z^{\tau+1} = v, W_1^1 = z_1^1,\dots,W_N^1 = z_N^1,\dots,W_1^{K_1} = z_1^{K_1},\dots W_N^{K_N} = z_N^{K_N} \mid\\
& Q_1^{\tau} = e_1,\dots,Q_n^{\tau} = e_n, R_1^{\tau} = g_1,\dots,R_n^{\tau} = g_n, Y^{\tau} = k, Z^{\tau} = u,\\
& W_1^1 = y_1^1,\dots,W_N^1 = y_N^1,\dots,W_1^{K_1} = y_1^{K_1},\dots W_N^{K_N} = y_N^{K_N})
\end{aligned}
\tag{B.6}
$$

- Transition probability for a system state extended by the remaining time of the time window B:

$$
\begin{aligned}
p_{xy} =& p_{(e_1,e_2,\dots,e_n,g_1,g_2,\dots,g_n,k,u,o),(f_1,f_2,\dots,f_n,h_1,h_2,\dots,h_n,l,v,p)} \\
=& P(\Lambda^{\tau+1} = y \mid \Lambda^{\tau} = x) \\
=& P(Q_1^{\tau+1} = f_1, \dots, Q_n^{\tau+1} = f_n, R_1^{\tau+1} = h_1, \dots, R_n^{\tau+1} = h_n, Y^{\tau+1} = l, \\
& Z^{\tau+1} = v, B^{\tau+1} = p \mid Q_1^{\tau} = e_1, \dots, Q_n^{\tau} = e_n, R_1^{\tau} = g_1, \dots, R_n^{\tau} = g_n, \\
& Y^{\tau} = k, Z^{\tau} = u, B^{\tau+1} = o)
\end{aligned}
\tag{B.7}
$$

- Transition probability for a system state extended by the remaining number of customers G:

$$
\begin{aligned}
p_{xy} =& p_{(e_1,e_2,\dots,e_n,g_1,g_2,\dots,g_n,k,u,q),(f_1,f_2,\dots,f_n,h_1,h_2,\dots,h_n,l,v,r)} \\
=& P(\Lambda^{\tau+1} = y \mid \Lambda^{\tau} = x) \\
=& P(Q_1^{\tau+1} = f_1, \dots, Q_n^{\tau+1} = f_n, R_1^{\tau+1} = h_1, \dots, R_n^{\tau+1} = h_n, Y^{\tau+1} = l, \\
& Z^{\tau+1} = v, G^{\tau+1} = r \mid Q_1^{\tau} = e_1, \dots, Q_n^{\tau} = e_n, R_1^{\tau} = g_1, \dots, R_n^{\tau} = g_n, \\
& Y^{\tau} = k, Z^{\tau} = u, G^{\tau+1} = q)
\end{aligned}
\tag{B.8}
$$

- Transition probability for a system state extended by the sink of the first customer O_i in the queue $i = \{1, 2, \cdots, N\}$:

$$
\begin{aligned}
p_{xy} =& p_{(e_1,e_2,\dots,e_n,g_1,g_2,\dots,g_n,k,u,m),(f_1,f_2,\dots,f_n,h_1,h_2,\dots,h_n,l,v,n)} \\
=& P(\Lambda^{\tau+1} = y \mid \Lambda^{\tau} = x) \\
=& P(Q_1^{\tau+1} = f_1, \dots, Q_n^{\tau+1} = f_n, R_1^{\tau+1} = h_1, \dots, R_n^{\tau+1} = h_n, Y^{\tau+1} = l, \\
& Z^{\tau+1} = v, O_1^{\tau+1} = n_1, \dots, O_N^{\tau+1} = n_N \mid Q_1^{\tau} = e_1, \dots, Q_n^{\tau} = e_n, \\
& R_1^{\tau} = g_1, \dots, R_n^{\tau} = g_n, Y^{\tau} = k, Z^{\tau} = u, \\
& O_1^{\tau+1} = m_1, \dots, O_N^{\tau+1} = m_N)
\end{aligned}
\tag{B.9}
$$

- Transition probability for a system state extended by the last departure time L_j per sink $j \in \{1,...,M\}$

$$
\begin{aligned}
p_{xy} =& p_{(e_1,e_2,...,e_n,g_1,g_2,...,g_n,k,u,s),(f_1,f_2,...,f_n,h_1,h_2,...,h_n,l,v,t)} \\
=& P(\Lambda^{\tau+1} = y \mid \Lambda^{\tau} = x) \\
=& P(Q_1^{\tau+1} = f_1, \ldots, Q_n^{\tau+1} = f_n, R_1^{\tau+1} = h_1, \ldots, R_n^{\tau+1} = h_n, Y^{\tau+1} = l, \\
& Z^{\tau+1} = v, L_1^{\tau+1} = t_1, \ldots, L_M^{\tau+1} = t_M \, | Q_1^{\tau} = e_1, \ldots, Q_n^{\tau} = e_n, \\
& R_1^{\tau} = g_1, \ldots, R_n^{\tau} = g_n, Y^{\tau} = k, Z^{\tau} = u, \\
& L_1^{\tau+1} = s_1, \ldots, L_M^{\tau+1} = s_M)
\end{aligned}
\tag{B.10}
$$

B.3 Additional Equations for the Calculation of the Transition Probability for the Extended System States

- Calculation of the transition probability of queue lengths Q_i, residual interarrival time R_i and waiting time of a customer W_i^b at position $b = \{1,2,\cdots,K_i\}$ of a queue $i = \{1,2,\cdots,N\}$:

$$
\begin{aligned}
&P_i(Q_i^{t+1}=e_i^{t+1}, R_i^{t+1}=g_i^{t+1}, W_i^{b^{t+1}}=y_i^{b^{t+1}} \mid Q_i^t=e_i^t, R_i^t=g_i^t, W_i^{b+\Theta^t}=y_i^{b+\Theta^t}, T^t=\Delta^t)\\
&=\begin{cases}
1 & b > e_i^{t+1} = e_i^t-\Theta,\ \ g_i^t > \Delta^t,\ \ g_i^{t+1} = g_i^t-\Delta^t,\ \ y_i^{b^{t+1}} = 0\\
1 & b \le e_i^{t+1} = e_i^t-\Theta,\ \ g_i^t > \Delta^t,\ \ g_i^{t+1} = g_i^t-\Delta^t,\ \ y_i^{b^{t+1}} = y_i^{b+\Theta^t}+\Delta^t\\
a_{i,\Delta^t-g_i^t+g_i^{t+1}} & K \ge b > e_i^{t+1} = e_i^t-\Theta+1,\ \ g_i^t \le \Delta^t,\ \ y_i^{b^{t+1}} = 0\\
a_{i,\Delta^t-g_i^t+g_i^{t+1}} & K > e_i^{t+1} = e_i^t-\Theta+1 > b,\ \ g_i^t \le \Delta^t,\ \ y_i^{b^{t+1}} = y_i^{b+\Theta^t}+\Delta^t\\
a_{i,\Delta^t-g_i^t+g_i^{t+1}} & K > e_i^{t+1} = b = e_i^t-\Theta+1,\ \ g_i^t \le \Delta^t,\ \ y_i^{b^{t+1}} = \Delta^t-g_i^t\\
\sum_{m=0}^{m_{max}} a_{i,\Delta^t-g_i^t-m}^{\otimes e_i^{t+1}-e_i^t-\Theta-2} \cdot a_{i,m+g_i^{t+1}} & K \ge b > e_i^{t+1} > e_i^t-\Theta+1,\ \ g_i^t \le \Delta^t,\ \ y_i^{b^{t+1}} = 0\\
\sum_{m=0}^{m_{max}} a_{i,\Delta^t-g_i^t-m}^{\otimes e_i^{t+1}-e_i^t-\Theta-2} \cdot a_{i,m+g_i^{t+1}} & K > e_i^{t+1} > e_i^t-\Theta+1 > b,\ \ g_i^t \le \Delta^t,\ \ y_i^{b^{t+1}} = y_i^{b+\Theta^t}+\Delta^t\\
\sum_{m=0}^{m_{max}} a_{i,\Delta^t-g_i^t-m}^{\otimes e_i^{t+1}-e_i^t-\Theta-2} \cdot a_{i,m+g_i^{t+1}} & K > e_i^{t+1} > b = e_i^t-\Theta+1,\ \ g_i^t \le \Delta^t,\ \ y_i^{b^{t+1}} = \Delta^t-g_i^t\\
a_{i,\Delta^t-g_i^t-y_i^{b^{t+1}}}^{\otimes b-e_i^t+\Theta-2} \cdot a_{i,y_i^{b^{t+1}}+g_i^{t+1}} & K > e_i^{t+1} = b > e_i^t-\Theta+1,\ \ g_i^t \le \Delta^t\\
a_{i,\Delta^t-g_i^t-y_i^{b^{t+1}}}^{\otimes b-e_i^t+\Theta-2} \sum_{l=0}^{l_{max}} a_{i,y_i^{b^{t+1}}-l}^{\otimes e_i^{t+1}-b-1} \cdot a_{i,l+g_i^{t+1}} & K > e_i^{t+1} > b > e_i^t-\Theta+1,\ \ g_i^t \le \Delta^t\\
a_{i,\Delta^t-g_i^t+g_i^{t+1}} + \sum_{m=0}^{m_{max}}\sum_{n=0}^{n_{max}} a_{i,\Delta^t-g_i^t-m}^{\otimes n} \cdot a_{i,m+g_i^{t+1}} & b \le e_i^t-\Theta \le e_i^{t+1} = K \le e_i^t-\Theta+1,\ \ y_i^{b^{t+1}} = y_i^{b+\Theta^t}+\Delta^t\\
a_{i,\Delta^t-g_i^t+g_i^{t+1}} + \sum_{m=0}^{m_{max}}\sum_{n=0}^{n_{max}} a_{i,\Delta^t-g_i^t-m}^{\otimes n} \cdot a_{i,m+g_i^{t+1}} & e_i^t-\Theta \le e_i^{t+1} = K \le b = e_i^t-\Theta+1,\ \ g_i^t \le \Delta^t,\ \ y_i^{b^{t+1}} = \Delta^t-g_i^t\\
\sum_{m=0}^{m_{max}}\sum_{o=0}^{o_{max}} a_{i,\Delta^t-g_i^t-m}^{\otimes e_i^{t+1}-e_i^t+\Theta+o-2} \cdot a_{i,m+g_i^{t+1}} & K = e_i^{t+1} > e_i^t-\Theta+1 > b,\ \ g_i^t \le \Delta^t,\ \ y_i^{b^{t+1}} = y_i^{b+\Theta^t}+\Delta^t\\
\sum_{m=0}^{m_{max}}\sum_{n=0}^{n_{max}} a_{i,\Delta^t-g_i^t-m}^{\otimes e_i^{t+1}-e_i^t+\Theta+n-2} \cdot a_{i,m+g_i^{t+1}} & K = e_i^{t+1} > b = e_i^t-\Theta+1,\ \ g_i^t \le \Delta^t,\ \ y_i^{b^{t+1}} = \Delta^t-g_i^t\\
a_{i,\Delta^t-g_i^t+y_i^{b^{t+1}}} \cdot \left(a_{i,y_i^{b^{t+1}}+g_i^{t+1}} + \sum_{l=0}^{l_{max}}\sum_{n=0}^{n_{max}} a_{i,y_i^{b^{t+1}}-l}^{\otimes n} \cdot a_{i,l+g_i^{t+1}}\right) & K = e_i^{t+1} > e_i^t-\Theta+1,\ \ g_i^t \le \Delta^t\\
a_{i,\Delta^t-g_i^t+y_i^{b^{t+1}}}^{\otimes b-e_i^t+\Theta-2} \cdot \sum_{l=0}^{l_{max}}\sum_{o=0}^{o_{max}} a_{i,y_i^{b^{t+1}}-l}^{\otimes e_i^{t+1}-b+o-1} \cdot a_{i,l+g_i^{t+1}} & K = e_i^{t+1} > b > e_i^t-\Theta+1,\ \ g_i^t \le \Delta^t\\
0 & \text{otherwise}
\end{cases}
\end{aligned}
\tag{B.11}
$$

- Calculation of the upper limits of the equation (B.11):

$$\begin{aligned} l_{max} &= \Delta^t - y_i^{b^{t+1}} - 1, \quad m_{max} = \Delta^t - g_i^t - 1 \\ n_{max} &= \left\lceil \frac{\Delta^t}{a_{i,min}} \right\rceil \\ o_{max} &= \left\lceil \frac{\Delta^t}{a_{i,min}} \right\rceil - e_i^{t+1} + e_i^t - \Theta \end{aligned} \tag{B.12}$$

- Calculation of the transition probability of queue lengths $Q_1,\ldots,Q_N$, residual interarrival time $R_1,\ldots,R_N$ and waiting time of a customer $W_1^1,\ldots,W_1^{K_i},\ldots,W_N^1,\ldots W_N^{K_i}$ for all positions and queues:

$$\begin{aligned} p^*_{x^{t+1}} =& P(\Lambda^{t+1} = x^{t+1} \mid \Lambda^t = x^t, T^t = \Delta^t) \\ =& P(Q_1^{t+1} = f_1,\ldots,Q_n^{t+1} = f_n, R_1^{t+1} = h_1,\ldots,R_n^{t+1} = h_n, \\ & W_1^1 = z_1^1,\ldots,W_N^1 = z_N^1,\ldots,W_1^{K_1} = z_1^{K_1},\ldots W_N^{K_N} = z_N^{K_N} \mid \\ & Q_1^t = e_1,\ldots,Q_n^t = e_n, R_1^t = g_1,\ldots,R_n^t = g_n, \\ & W_1^1 = y_1^1,\ldots,W_N^1 = y_N^1,\ldots,W_1^{K_1} = y_1^{K_1},\ldots W_N^{K_N} = y_N^{K_N}, \\ & T^t = \Delta^t) \\ =& \prod_{b=1}^{K}\prod_{i=1}^{N} P_i(Q_i^{t+1} = e_i^{t+1}, R_i^{t+1} = g_i^{t+1}, W_i^{b^{t+1}} = y_i^{b^{t+1}} \mid \\ & Q_i^t = e_i^t, R_i^t = g_i^t, W_i^{b+\Theta^t} = y_i^{b+\Theta^t}, T^t = \Delta^t) \end{aligned} \tag{B.13}$$

- Calculation of the transition probability of the remaining time of the time window B:

$$p_{1,additional} = \begin{cases} 1 & p_{continue} = 1, \quad p = o - \Delta \\ 1 & p_{continue} = 0, \quad p = TW_l \\ 0 & \text{otherwise} \end{cases} \tag{B.14}$$

- Calculation of the transition probability of the remaining number of customers G:

$$p_{2,additional} = \begin{cases} 1 & p_{continue} = 1, \quad RT_2 \cap \{3,4\} \neq \emptyset, \quad r = q-1 \\ 1 & p_{continue} = 0, \quad RT_2 \in \{3\}, \quad RT_2 \notin \{4\}, \quad r = MN_l \\ 1 & p_{continue} = 0, \quad RT_2 \notin \{3\}, \quad RT_2 \in \{4\}, \quad r = \max\{f_l, 1\} \\ 1 & p_{continue} = 0, \quad RT_2 \in \{3,4\}, \quad r = \max\{\min\{MN_l, f_l\}, 1\} \\ 0 & \text{otherwise} \end{cases} \tag{B.15}$$

- Calculation of the transition probability of the sink of the first customer O_i in the queue $i = \{1, 2, \cdots, N\}$:

$$P\left(G_i^{\tau+1} = n_i \mid Y^\tau = k, G_i^\tau = m_i\right) = \begin{cases} 1 & i \neq k, \quad n_i = m_i \\ \hat{p}_{i,n_i} & i = k \\ 0 & \text{otherwise} \end{cases} \tag{B.16}$$

- Calculation of the transition probability of the sink of the first customer $O_1, \ldots, O_N$ for all queues:

$$p_{3,additional} = \prod_{i=1}^{N} P\left(G_i^{\tau+1} = n_i \mid Y^\tau = k, G_i^\tau = m_i\right) \tag{B.17}$$

- Calculation of the transition probability of the last departure time L_j per sink $j = \{1, \cdots, M\}$:

$$P\left(L_j^{\tau+1} = t_j \mid L_j^\tau = s_j, Z^{\tau+1} = v, T^4 = \Delta^4, T^5 = \Delta^5, T = \Delta\right) = \begin{cases} 1 & j \neq v, \quad t_j = s_j + \Delta \\ 1 & j = v, \quad t_j = \Delta^4 + \Delta^5 \\ 0 & \text{otherwise} \end{cases} \tag{B.18}$$

- Calculation of the transition probability of the last departure time $L_1, \ldots, L_M$ for all sinks:

$$p_{4,additional} = \prod_{j=1}^{M} P\left(L_j^{\tau+1} = t_j \mid L_j^{\tau} = s_j, Z^{\tau+1} = v, T^4 = \Delta^4, T^5 = \Delta^5, T = \Delta\right) \tag{B.19}$$

B.4 Calculation of the Transition Probability for the Extended System States

- Calculation of the transition probability for a system state extended by the waiting time of a customer W_i^b at position $b = \{1, 2, \cdots, K_i\}$ in the queue $i = \{1, 2, \cdots, N\}$:

$$\begin{aligned} p_{xy} =& P(\Lambda^{\tau+1} = y \mid \Lambda^{\tau} = x) = P(\Lambda^{\tau+1} = x^6 \mid \Lambda^{\tau} = x) \\ =& \sum_{\Delta^1=0}^{\Delta^1_{max}} p_{\Delta^1} \cdot \sum_{x^2=0}^{x_{max}} p^*_{x^2} \cdot \sum_{\Delta^2=0}^{\Delta^2_{max}} p_{\Delta^2} \cdot \sum_{x^3=0}^{x_{max}} p^*_{x^3} \cdot \sum_{\Delta^3=0}^{\Delta^3_{max}} p_{\Delta^3} \cdot \sum_{x^4=0}^{x_{max}} p^*_{x^4} \\ & \cdot \sum_{\Delta^4=0}^{\Delta^4_{max}} p_{\Delta^4} \cdot \sum_{x^5=0}^{x_{max}} p^*_{x^5} \cdot \sum_{\ddot{\Delta}=0}^{\ddot{\Delta}_{max}} p_{\ddot{\Delta}} \cdot \sum_{\Delta^5=0}^{\Delta^5_{max}} p_{\Delta^5} \cdot p^*_{x^6} \cdot p_{next} \cdot p_{transition} \end{aligned} \tag{B.20}$$

- Calculation of the transition probability for a system state extended by the remaining time of the time window B:

$$\begin{aligned} p_{xy} =& P(\Lambda^{\tau+1} = y \mid \Lambda^{\tau} = x) = P(\Lambda^{\tau+1} = x^6 \mid \Lambda^{\tau} = x) \\ =& \sum_{\Delta^1=0}^{\Delta^1_{max}} p_{\Delta^1} \cdot \sum_{x^2=0}^{x_{max}} p_{x^2} \cdot \sum_{\Delta^2=0}^{\Delta^2_{max}} p_{\Delta^2} \cdot \sum_{x^3=0}^{x_{max}} p_{x^3} \cdot \sum_{\Delta^3=0}^{\Delta^3_{max}} p_{\Delta^3} \cdot \sum_{x^4=0}^{x_{max}} p_{x^4} \\ & \cdot \sum_{\Delta^4=0}^{\Delta^4_{max}} p_{\Delta^4} \cdot \sum_{x^5=0}^{x_{max}} p_{x^5} \cdot \sum_{\ddot{\Delta}=0}^{\ddot{\Delta}_{max}} p_{\ddot{\Delta}} \cdot \sum_{\Delta^5=0}^{\Delta^5_{max}} p_{\Delta^5} \cdot p_{x^6} \cdot p_{next} \cdot p_{transition} \\ & \cdot p_{1,additional} \end{aligned} \tag{B.21}$$

- Calculation of the transition probability for a system state extended by the remaining number of customers G:

$$\begin{aligned} p_{xy} =& P(\Lambda^{\tau+1} = y \mid \Lambda^{\tau} = x) = P(\Lambda^{\tau+1} = x^6 \mid \Lambda^{\tau} = x) \\ =& \sum_{\Delta^1=0}^{\Delta^1_{max}} p_{\Delta^1} \cdot \sum_{x^2=0}^{x_{max}} p_{x^2} \cdot \sum_{\Delta^2=0}^{\Delta^2_{max}} p_{\Delta^2} \cdot \sum_{x^3=0}^{x_{max}} p_{x^3} \cdot \sum_{\Delta^3=0}^{\Delta^3_{max}} p_{\Delta^3} \cdot \sum_{x^4=0}^{x_{max}} p_{x^4} \\ & \cdot \sum_{\Delta^4=0}^{\Delta^4_{max}} p_{\Delta^4} \cdot \sum_{x^5=0}^{x_{max}} p_{x^5} \cdot \sum_{\ddot{\Delta}=0}^{\ddot{\Delta}_{max}} p_{\ddot{\Delta}} \cdot \sum_{\Delta^5=0}^{\Delta^5_{max}} p_{\Delta^5} \cdot p_{x^6} \cdot p_{next} \cdot p_{transition} \\ & \cdot p_{2,additional} \end{aligned} \tag{B.22}$$

- Calculation of the transition probability for a system state extended by the sink of the first customer O_i in the queue $i = \{1, 2, \cdots, N\}$:

$$\begin{aligned} p_{xy} =& P(\Lambda^{\tau+1} = y \mid \Lambda^{\tau} = x) = P(\Lambda^{\tau+1} = x^6 \mid \Lambda^{\tau} = x) \\ =& \sum_{\Delta^1=0}^{\Delta^1_{max}} p_{\Delta^1} \cdot \sum_{x^2=0}^{x_{max}} p_{x^2} \cdot \sum_{\Delta^2=0}^{\Delta^2_{max}} p_{\Delta^2} \cdot \sum_{x^3=0}^{x_{max}} p_{x^3} \cdot \sum_{\Delta^3=0}^{\Delta^3_{max}} p_{\Delta^3} \cdot \sum_{x^4=0}^{x_{max}} p_{x^4} \\ & \cdot \sum_{\Delta^4=0}^{\Delta^4_{max}} p_{\Delta^4} \cdot \sum_{x^5=0}^{x_{max}} p_{x^5} \cdot \sum_{\ddot{\Delta}=0}^{\ddot{\Delta}_{max}} p_{\ddot{\Delta}} \cdot \sum_{\Delta^5=0}^{\Delta^5_{max}} p_{\Delta^5} \cdot p_{x^6} \cdot p_{next} \cdot p_{transition} \\ & \cdot p_{3,additional} \end{aligned} \tag{B.23}$$

- Calculation of the transition probability for a system state extended by the last departure time L_j per sink $j \in \{1, ..., M\}$

$$\begin{aligned} p_{xy} =& P(\Lambda^{\tau+1} = y \mid \Lambda^{\tau} = x) = P(\Lambda^{\tau+1} = x^6 \mid \Lambda^{\tau} = x) \\ =& \sum_{\Delta^1=0}^{\Delta^1_{max}} p_{\Delta^1} \cdot \sum_{x^2=0}^{x_{max}} p_{x^2} \cdot \sum_{\Delta^2=0}^{\Delta^2_{max}} p_{\Delta^2} \cdot \sum_{x^3=0}^{x_{max}} p_{x^3} \cdot \sum_{\Delta^3=0}^{\Delta^3_{max}} p_{\Delta^3} \cdot \sum_{x^4=0}^{x_{max}} p_{x^4} \\ & \cdot \sum_{\Delta^4=0}^{\Delta^4_{max}} p_{\Delta^4} \cdot \sum_{x^5=0}^{x_{max}} p_{x^5} \cdot \sum_{\ddot{\Delta}=0}^{\ddot{\Delta}_{max}} p_{\ddot{\Delta}} \cdot \sum_{\Delta^5=0}^{\Delta^5_{max}} p_{\Delta^5} \cdot p_{x^6} \cdot p_{next} \cdot p_{transition} \\ & \cdot p_{4,additional} \end{aligned} \tag{B.24}$$

B.5 Calculation of the Expected Value of the Switching Time from a Sink to a Queue with Different Distributions over the Queues and Sinks

- Calculation of the expected value of the switching time $E(C_{ij}^{(1),(1)})$ from a sink $j \in \{1,\ldots,M\}$ to a queue $i \in \{1,\ldots,N\}$ with a uniform distribution over the queues and the sinks:

$$E(C_{ji}^{(1),(1)}) = \begin{cases} E(\tilde{C}) & BV_2 = 0 \\ \frac{N}{N-1} \cdot E(\tilde{C}) & BV_2 = 1, \quad i \neq j \\ 0 & BV_2 = 1, \quad i = j \end{cases} \tag{B.25}$$

- Calculation of the expected value of the switching time $E(C_{ij}^{(2),(1)})$ from a sink $j \in \{1,\ldots,M\}$ to a queue $i \in \{1,\ldots,N\}$ with a unequal distribution over the queues with the gradient b_{C_i} and the constant y_{C_i} and a uniform distribution over the sinks:

$$E(C_{ji}^{(2),(1)}) = \begin{cases} (i-1) \cdot b_{C_i} + y_{C_i} & BV_2 = 0 \\ (i-1) \cdot b_{C_i} + y_{C_i} & BV_2 = 1, \quad i \neq j \\ 0 & BV_2 = 1, \quad i = j \end{cases} \tag{B.26}$$

- Calculation of the expected value of the switching time $E(C_{ij}^{(1),(2)})$ from a sink $j \in \{1,\ldots,M\}$ to a queue $i \in \{1,\ldots,N\}$ with a uniform distribution over the queues and a unequal distribution over the sinks with the gradient b_{C_j} and the constant y_{C_j}:

$$E(C_{ji}^{(1),(2)}) = \begin{cases} (j-1) \cdot b_{C_j} + y_{C_j} & BV_2 = 0 \\ (j-1) \cdot b_{C_j} + y_{C_j} & BV_2 = 1, \quad i \neq j \\ 0 & BV_2 = 1, \quad i = j \end{cases} \tag{B.27}$$

- Calculation of the expected value of the switching time $E(C_{ij}^{(2),(2)})$ from a sink $j \in \{1,\ldots,M\}$ to a queue $i \in \{1,\ldots,N\}$ with a unequal distribution over the queues and the sinks:

$$E(C_{ji}^{(2),(2)}) = \frac{E(C_{ji}^{(2),(1)}) + E(C_{ji}^{(1),(2)})}{2} \tag{B.28}$$

C Parameter Configurations

In the following, tables of the parameter configurations for the validation of the simulation model (see Chapter 6) and for the numerical evaluation (see Chapter 7) are presented.

C.1 Table of the Parameter Configurations for Validation

No.	N/M	K_i	p_{ij}		a_i		b_{i1}		b_{i2}		c_{j1}		c_{j2}		PN_i	$\tilde{p}_l$	TW_i	MN_i	LV_1	LV_2
1					8	9	2	3	1	2	1	2	1	2						
	2	3	1.00	0.00	0.37	0.63	0.70	0.30	0.00	0.00	0.53	0.47	0.53	0.47	2	0.57	5	2	2	8
	1	5	1.00	0.00	0.20	0.80	0.24	0.76	0.00	0.00	0.00	0.00	0.00	0.00	1	0.43	7	2	2	9
2					6	7	2	3	1	3	0	1	0	2						
	2	5	0.39	0.61	0.26	0.74	0.37	0.63	0.74	0.26	0.10	0.90	0.79	0.21	1	0.42	6	3	4	5
	2	5	0.61	0.39	0.80	0.20	0.50	0.50	0.26	0.74	0.27	0.73	0.28	0.72	2	0.58	5	2	3	6
3					6	7	1	3	1	3	0	2	1	2						
	2	2	1.00	0.00	0.02	0.98	0.80	0.20	0.00	0.00	0.04	0.96	0.31	0.69	2	0.97	6	4	2	8
	1	4	1.00	0.00	0.87	0.13	0.76	0.24	0.00	0.00	0.00	0.00	0.00	0.00	1	0.03	5	3	2	9
4					6	9	1	2	2	3	1	2	1	2						
	2	4	0.24	0.76	0.69	0.31	0.86	0.14	0.81	0.19	0.17	0.83	0.35	0.65	1	0.22	10	2	3	6
	2	2	0.72	0.28	0.74	0.26	0.94	0.06	0.32	0.68	0.75	0.25	0.30	0.70	2	0.78	5	3	2	4
5					5	7	1	3	2	3	1	2	0	1						
	2	4	0.79	0.21	0.17	0.83	0.88	0.12	0.74	0.26	0.48	0.52	0.96	0.04	2	0.52	5	3	4	8
	2	2	0.13	0.87	0.88	0.12	0.61	0.39	0.51	0.49	0.90	0.10	0.48	0.52	1	0.48	4	3	2	5
6					5	8	1	2	2	3	0	2	0	1						
	2	4	1.00	0.00	1.00	0.00	0.04	0.96	0.00	0.00	0.11	0.89	0.82	0.18	2	0.34	10	3	2	4
	1	4	1.00	0.00	0.36	0.64	0.88	0.12	0.00	0.00	0.00	0.00	0.00	0.00	1	0.66	7	2	2	5
7					6	8	2	3	1	3	0	2	0	2						
	2	5	0.76	0.24	0.78	0.22	0.98	0.02	0.64	0.36	0.12	0.88	0.74	0.26	1	0.32	4	2	3	4
	2	4	0.74	0.26	0.58	0.42	0.73	0.27	0.38	0.62	0.88	0.12	0.32	0.68	2	0.68	5	3	3	7
8					4	9	2	3	1	3	0	1	1	2						
	2	2	0.60	0.40	0.48	0.52	0.40	0.60	0.93	0.07	0.07	0.93	0.44	0.56	1	0.04	7	2	2	5
	2	5	0.74	0.26	0.17	0.83	0.75	0.25	0.31	0.69	0.05	0.95	0.06	0.94	2	0.96	10	2	5	4

No.	N/M	K_i	p_{ij}		a_i		b_{i1}		b_{i2}		c_{j1}		c_{j2}		PN_i	$\tilde{p}_l$	TW_i	MN_i	LV_1	LV_2
9					7	8	2	3	1	2	1	2	0	2						
	2	3	1.00	0.00	0.19	0.81	0.72	0.28	0.00	0.00	0.40	0.60	0.33	0.67	1	0.99	8	2	2	7
	1	5	1.00	0.00	0.25	0.75	0.47	0.53	0.00	0.00	0.00	0.00	0.00	0.00	2	0.01	5	2	3	4
10					6	8	2	3	2	3	1	2	0	1						
	2	4	0.07	0.93	0.51	0.49	0.60	0.40	0.51	0.49	0.92	0.08	0.84	0.16	2	0.08	8	2	4	4
	2	3	0.37	0.63	0.57	0.43	0.49	0.51	0.64	0.36	0.22	0.78	0.81	0.19	1	0.92	10	2	3	6
11					8	10	2	3	1	3	0	2	0	1						
	2	4	0.93	0.07	0.79	0.21	0.98	0.02	0.84	0.16	0.34	0.66	0.56	0.44	1	0.65	7	3	4	10
	2	4	0.79	0.21	0.86	0.14	0.52	0.48	0.78	0.22	0.83	0.17	0.01	0.99	2	0.35	8	4	3	7
12					8	10	1	3	1	3	1	2	1	2						
	2	3	1.00	0.00	0.45	0.55	0.90	0.10	0.00	0.00	0.71	0.29	0.17	0.83	1	0.41	6	3	2	4
	1	4	1.00	0.00	0.07	0.93	0.27	0.73	0.00	0.00	0.00	0.00	0.00	0.00	2	0.59	4	2	3	7
13					8	10	1	2	2	3	0	2	1	2						
	2	5	1.00	0.00	0.08	0.92	0.04	0.96	0.00	0.00	0.06	0.94	0.61	0.39	2	0.47	9	4	2	7
	1	3	1.00	0.00	0.27	0.73	0.35	0.65	0.00	0.00	0.00	0.00	0.00	0.00	1	0.53	10	3	2	9
14					8	10	1	3	1	3	1	2	1	2						
	2	2	0.44	0.56	0.28	0.72	0.23	0.77	0.30	0.70	0.07	0.93	0.62	0.38	2	0.62	7	2	2	6
	2	2	0.44	0.56	0.14	0.86	0.08	0.92	0.11	0.89	0.88	0.12	0.81	0.19	1	0.38	5	4	2	10
15					7	10	2	3	2	3	0	2	1	2						
	2	2	1.00	0.00	0.77	0.23	0.99	0.01	0.00	0.00	0.79	0.21	0.08	0.92	1	0.37	10	2	2	7
	1	2	1.00	0.00	0.06	0.94	0.24	0.76	0.00	0.00	0.00	0.00	0.00	0.00	2	0.63	10	4	2	5
16					7	10	1	3	1	2	0	2	1	2						
	2	2	1.00	0.00	0.45	0.55	0.58	0.42	0.00	0.00	0.07	0.93	0.30	0.70	1	0.58	4	3	2	7
	1	2	1.00	0.00	0.87	0.13	0.66	0.34	0.00	0.00	0.00	0.00	0.00	0.00	2	0.42	4	4	2	5

No.	N/M	K_i	p_{ij}		a_i		b_{i1}		b_{i2}		c_{j1}		c_{j2}		PN_i	$\tilde{p}_l$	TW_i	MN_i	LV_1	LV_2
17					7	8	2	3	2	3	0	1	1	2						
	2	5	1.00	0.00	0.66	0.34	1.00	0.00	0.00	0.00	0.76	0.24	0.87	0.13	2	0.89	5	4	2	8
	1	5	1.00	0.00	0.19	0.81	0.19	0.81	0.00	0.00	0.00	0.00	0.00	0.00	1	0.11	10	4	2	9
18					4	10	1	2	2	3	0	2	0	2						
	2	5	1.00	0.00	0.74	0.26	0.25	0.75	0.00	0.00	0.57	0.43	0.56	0.44	1	0.27	7	4	5	9
	1	3	1.00	0.00	0.32	0.68	0.77	0.23	0.00	0.00	0.00	0.00	0.00	0.00	2	0.73	8	3	3	10
19					7	8	1	3	1	3	1	2	0	2						
	2	5	0.04	0.96	0.17	0.83	0.86	0.14	0.15	0.85	0.93	0.07	0.87	0.13	1	0.07	5	2	2	9
	2	2	0.58	0.42	0.26	0.74	0.55	0.45	0.65	0.35	0.62	0.38	0.46	0.54	2	0.93	4	4	2	4
20					7	8	2	3	1	3	1	2	0	1						
	2	3	1.00	0.00	0.49	0.51	0.81	0.19	0.00	0.00	0.83	0.17	0.20	0.80	2	0.45	6	3	3	6
	1	3	1.00	0.00	0.06	0.94	0.36	0.64	0.00	0.00	0.00	0.00	0.00	0.00	1	0.55	10	3	3	10
21					9	10	2	3	1	2	1	2	0	1						
	2	5	0.96	0.04	0.52	0.48	0.69	0.31	0.58	0.42	0.65	0.35	0.20	0.80	2	0.95	6	4	4	4
	2	4	0.84	0.16	0.16	0.84	0.80	0.20	0.56	0.44	0.05	0.95	0.61	0.39	1	0.05	5	2	2	8
22					6	9	2	3	1	2	0	1	0	1						
	2	4	0.03	0.97	0.13	0.87	0.73	0.27	0.56	0.44	0.82	0.18	0.52	0.48	1	0.06	4	2	4	8
	2	2	0.58	0.42	0.97	0.03	0.78	0.22	0.14	0.86	0.73	0.27	0.90	0.10	2	0.94	10	2	2	4
23					4	10	1	2	1	2	0	1	0	1						
	2	5	0.54	0.46	0.58	0.42	0.71	0.29	0.87	0.13	0.50	0.50	0.50	0.50	2	0.06	10	2	2	5
	2	5	0.24	0.76	0.30	0.70	0.77	0.23	0.96	0.04	0.18	0.82	0.24	0.76	1	0.94	4	2	2	10
24					9	10	1	2	1	3	1	2	0	2						
	2	3	0.75	0.25	0.63	0.37	0.43	0.57	0.76	0.24	0.36	0.64	0.29	0.71	2	0.68	4	2	3	6
	2	5	0.03	0.97	0.20	0.80	0.60	0.40	0.19	0.81	0.25	0.75	0.93	0.07	1	0.32	8	4	2	4

No.	N/M	K_i	p_{ij}		a_i		b_{i1}		b_{i2}		c_{j1}		c_{j2}		PN_i	$\tilde{p}_l$	TW_i	MN_i	LV_1	LV_2
25					7	8	1	3	1	2	0	2	0	1						
	2	2	0.37	0.63	0.99	0.01	0.54	0.46	0.73	0.27	0.55	0.45	0.26	0.74	2	0.84	10	4	2	9
	2	5	0.92	0.08	0.09	0.91	0.57	0.43	0.96	0.04	0.67	0.33	0.68	0.32	1	0.16	5	4	3	8
26					7	9	2	3	2	3	0	1	0	1						
	2	4	1.00	0.00	0.29	0.71	0.59	0.41	0.00	0.00	0.95	0.05	0.83	0.17	1	0.71	10	3	2	5
	1	4	1.00	0.00	0.10	0.90	1.00	0.00	0.00	0.00	0.00	0.00	0.00	0.00	2	0.29	8	4	3	5
27					9	10	1	3	1	3	1	2	0	2						
	2	3	0.36	0.64	0.74	0.26	0.10	0.90	0.72	0.28	0.46	0.54	0.26	0.74	2	0.34	7	3	2	5
	2	4	0.87	0.13	0.83	0.17	0.66	0.34	0.15	0.85	0.86	0.14	0.66	0.34	1	0.66	6	4	3	10
28					9	10	2	3	1	3	0	2	0	1						
	2	3	1.00	0.00	0.84	0.16	0.61	0.39	0.00	0.00	0.27	0.73	0.94	0.06	1	0.82	9	3	3	7
	1	3	1.00	0.00	0.23	0.77	0.68	0.32	0.00	0.00	0.00	0.00	0.00	0.00	2	0.18	8	3	3	10
29					7	10	1	2	1	2	1	2	0	2						
	2	3	0.78	0.22	0.67	0.33	0.57	0.43	0.04	0.96	0.68	0.32	0.48	0.52	2	0.76	8	3	2	7
	2	4	0.64	0.36	0.09	0.91	0.18	0.82	0.53	0.47	0.23	0.77	0.60	0.40	1	0.24	9	3	2	9
30					9	10	2	3	1	2	0	2	0	1						
	2	5	0.57	0.43	0.59	0.41	0.11	0.89	0.18	0.82	0.88	0.12	0.71	0.29	1	0.40	8	3	3	4
	2	2	0.32	0.68	0.70	0.30	0.98	0.02	0.20	0.80	0.51	0.49	0.27	0.73	2	0.60	7	2	2	9
31					2	5	1	2	1	2	0	1	0	1						
	2	2	0.75	0.25	0.20	0.80	0.04	0.96	0.12	0.88	0.98	0.02	0.54	0.46	1	0.56	4	2	1	3
	2	3	0.12	0.88	0.06	0.94	0.32	0.68	0.10	0.90	0.69	0.31	0.94	0.06	2	0.44	2	3	2	3
32					4	5	1	2	1	2	0	1	0	1						
	2	2	1.00	0.00	0.28	0.72	0.49	0.51	0.00	0.00	0.37	0.63	0.81	0.19	2	0.56	4	3	2	3
	1	3	1.00	0.00	0.52	0.48	0.11	0.89	0.00	0.00	0.00	0.00	0.00	0.00	1	0.44	4	2	3	2

No.	N/M	K_i	p_{ij}		a_i		b_{i1}		b_{i2}		c_{j1}		c_{j2}		PN_i	$\tilde{p}_l$	TW_i	MN_i	LV_1	LV_2
33					3	4	1	2	1	2	0	1	0	1						
	2	3	1.00	0.00	0.50	0.50	0.67	0.33	0.00	0.00	0.42	0.58	0.65	0.35	2	0.12	3	3	3	2
	1	2	1.00	0.00	0.47	0.53	0.93	0.07	0.00	0.00	0.00	0.00	0.00	0.00	1	0.88	2	2	1	4
34					4	5	1	2	1	2	0	1	0	1						
	2	2	1.00	0.00	0.82	0.18	0.23	0.77	0.00	0.00	0.24	0.76	0.80	0.20	1	1.00	3	2	2	2
	1	2	1.00	0.00	0.70	0.30	0.78	0.22	0.00	0.00	0.00	0.00	0.00	0.00	2	0.00	3	1	1	4
35					3	5	1	2	1	2	0	1	0	1						
	2	2	0.52	0.48	0.52	0.48	0.88	0.12	0.30	0.70	0.58	0.42	0.59	0.41	2	0.89	4	2	2	3
	2	1	0.15	0.85	0.00	1.00	0.49	0.51	0.47	0.53	0.14	0.86	0.26	0.74	1	0.11	4	2	1	2
36					4	5	1	2	1	2	0	1	0	1						
	2	3	0.02	0.98	0.64	0.36	0.36	0.64	0.38	0.62	0.73	0.27	0.54	0.46	2	0.85	2	3	3	4
	2	1	0.40	0.60	0.36	0.64	0.27	0.73	0.79	0.21	0.71	0.29	0.20	0.80	1	0.15	4	1	1	2
37					3	5	1	2	1	2	0	1	0	1						
	2	1	0.27	0.73	0.18	0.82	0.75	0.25	0.25	0.75	0.32	0.68	0.09	0.91	2	0.51	2	2	1	3
	2	1	0.24	0.76	0.40	0.60	0.53	0.47	0.74	0.26	0.37	0.63	0.40	0.60	1	0.49	3	3	1	2
38					4	5	1	2	1	2	0	1	0	1						
	2	1	0.57	0.43	0.09	0.91	0.48	0.52	0.52	0.48	0.41	0.59	0.09	0.91	2	0.45	3	3	1	4
	2	1	0.56	0.44	0.04	0.96	0.21	0.79	0.37	0.63	0.40	0.60	0.45	0.55	1	0.55	3	2	1	4
39					3	4	1	2	1	2	0	1	0	1						
	2	1	0.75	0.25	0.13	0.87	0.68	0.32	0.27	0.73	0.71	0.29	0.83	0.17	1	0.83	2	1	1	3
	2	2	0.33	0.67	0.81	0.19	0.25	0.75	0.96	0.04	0.30	0.70	0.65	0.35	2	0.17	4	1	2	3
40					2	5	1	2	1	2	0	1	0	1						
	2	3	0.83	0.17	0.51	0.49	0.58	0.42	0.56	0.44	0.84	0.16	0.67	0.33	1	0.35	4	1	2	2
	2	2	0.67	0.33	0.39	0.61	0.26	0.74	0.32	0.68	0.81	0.19	0.65	0.35	2	0.65	4	3	1	3

No.	N/M	K_i	p_{ij}		a_i		b_{i1}		b_{i2}		c_{j1}		c_{j2}		PN_i	$\tilde{p}_l$	TW_i	MN_i	LV_1	LV_2
41					2	5	1	2	1	2	0	1	0	1						
	2	2	1.00	0.00	0.66	0.34	0.61	0.39	0.00	0.00	0.80	0.20	0.84	0.16	2	0.61	2	2	1	3
	1	1	1.00	0.00	0.48	0.52	0.85	0.15	0.00	0.00	0.00	0.00	0.00	0.00	1	0.39	4	3	1	3
42					4	5	1	2	1	2	0	1	0	1						
	2	3	0.26	0.74	0.07	0.93	0.44	0.56	0.31	0.69	0.26	0.74	0.04	0.96	2	0.49	4	1	3	2
	2	2	0.48	0.52	0.45	0.55	0.96	0.04	0.87	0.13	0.92	0.08	0.90	0.10	1	0.51	2	3	2	3
43					4	5	1	2	1	2	0	1	0	1						
	2	3	0.04	0.96	0.95	0.05	0.59	0.41	0.68	0.32	0.79	0.21	0.65	0.35	2	0.08	4	2	1	2
	2	1	0.71	0.29	0.39	0.61	0.35	0.65	0.14	0.86	0.66	0.34	0.82	0.18	1	0.92	2	3	1	3
44					1	5	1	2	1	2	0	1	0	1						
	2	1	0.30	0.70	0.36	0.64	0.82	0.18	0.57	0.43	0.37	0.63	0.60	0.40	2	0.54	2	3	1	2
	2	2	0.75	0.25	0.15	0.85	0.69	0.31	0.55	0.45	0.48	0.52	0.97	0.03	1	0.46	3	1	2	3
45					4	5	1	2	1	2	0	1	0	1						
	2	2	0.17	0.83	0.53	0.47	0.09	0.91	0.45	0.55	0.98	0.02	0.07	0.93	2	0.20	4	2	2	4
	2	2	0.19	0.81	0.14	0.86	0.74	0.26	0.29	0.71	0.82	0.18	0.63	0.37	1	0.80	2	2	2	2
46					4	5	1	2	1	2	0	1	0	1						
	2	1	1.00	0.00	0.02	0.98	0.74	0.26	0.00	0.00	0.53	0.47	0.81	0.19	1	0.09	4	3	1	2
	1	3	1.00	0.00	0.73	0.27	0.34	0.66	0.00	0.00	0.00	0.00	0.00	0.00	2	0.91	3	3	1	2
47					4	5	1	2	1	2	0	1	0	1						
	2	3	0.21	0.79	0.95	0.05	0.36	0.64	0.94	0.06	0.62	0.38	0.05	0.95	2	0.06	3	2	2	2
	2	2	0.83	0.17	0.81	0.19	0.83	0.17	0.66	0.34	0.70	0.30	0.14	0.86	1	0.94	2	3	2	4
48					3	5	1	2	1	2	0	1	0	1						
	2	1	0.90	0.10	0.23	0.77	0.83	0.17	0.02	0.98	0.24	0.76	0.83	0.17	2	0.46	3	3	1	3
	2	3	0.52	0.48	0.27	0.73	0.98	0.02	0.94	0.06	0.39	0.61	0.88	0.12	1	0.54	4	3	2	2

No.	N/M	K_i	p_{ij}		a_i		b_{i1}		b_{i2}		c_{j1}		c_{j2}		PN_i	$\tilde{p}_l$	TW_i	MN_i	LV_1	LV_2
49					2	5	1	2	1	2	0	1	0	1						
	2	2	0.09	0.91	0.30	0.70	0.34	0.66	0.83	0.17	0.75	0.25	0.87	0.13	2	0.22	3	2	2	3
	2	1	0.89	0.11	0.41	0.59	0.87	0.13	0.32	0.68	0.15	0.85	0.15	0.85	1	0.78	2	1	1	4
50					4	5	1	2	1	2	0	1	0	1						
	2	3	0.81	0.19	0.56	0.44	0.94	0.06	0.03	0.97	0.58	0.42	0.73	0.27	1	0.16	4	2	2	4
	2	3	0.11	0.89	0.35	0.65	0.15	0.85	0.67	0.33	0.90	0.10	0.12	0.88	2	0.84	3	1	3	2
51					4	5	1	2	1	2	0	1	0	1						
	2	2	0.66	0.34	0.20	0.80	0.88	0.12	0.94	0.06	0.68	0.32	0.05	0.95	1	0.58	3	2	1	3
	2	1	0.82	0.18	0.22	0.78	0.98	0.02	0.98	0.02	0.89	0.11	0.50	0.50	2	0.42	4	1	1	3
52					4	5	1	2	1	2	0	1	0	1						
	2	3	1.00	0.00	0.11	0.89	0.46	0.54	0.00	0.00	0.67	0.33	0.61	0.39	1	0.71	2	2	2	3
	1	2	1.00	0.00	0.07	0.93	0.91	0.09	0.00	0.00	0.00	0.00	0.00	0.00	2	0.29	2	1	1	4
53					4	5	1	2	1	2	0	1	0	1						
	2	3	0.96	0.04	0.52	0.48	0.94	0.06	0.15	0.85	0.74	0.26	0.98	0.02	1	0.81	3	2	2	3
	2	3	0.36	0.64	0.08	0.92	0.60	0.40	0.18	0.82	0.89	0.11	0.85	0.15	2	0.19	3	3	1	3
54					4	5	1	2	1	2	0	1	0	1						
	2	1	1.00	0.00	0.51	0.49	0.81	0.19	0.00	0.00	0.86	0.14	0.74	0.26	2	0.46	2	2	1	3
	1	1	1.00	0.00	0.22	0.78	0.55	0.45	0.00	0.00	0.00	0.00	0.00	0.00	1	0.54	2	3	1	2
55					4	5	1	2	1	2	0	1	0	1						
	2	1	0.53	0.47	0.19	0.81	0.49	0.51	0.89	0.11	0.83	0.17	0.99	0.01	1	0.02	3	2	1	2
	2	2	0.10	0.90	0.03	0.97	0.29	0.71	0.93	0.07	0.95	0.05	0.65	0.35	2	0.98	2	2	2	2
56					4	5	1	2	1	2	0	1	0	1						
	2	3	1.00	0.00	0.72	0.28	0.85	0.15	0.00	0.00	0.47	0.53	0.84	0.16	2	0.69	2	3	3	2
	1	3	1.00	0.00	0.44	0.56	0.92	0.08	0.00	0.00	0.00	0.00	0.00	0.00	1	0.31	2	2	1	4

No.	N/M	K_i	p_{ij}		a_i		b_{i1}		b_{i2}		c_{j1}		c_{j2}		PN_i	$\tilde{p}_l$	TW_i	MN_i	LV_1	LV_2
57					4	5	1	2	1	2	0	1	0	1						
	2	2	1.00	0.00	0.17	0.83	0.32	0.68	0.00	0.00	0.96	0.04	0.77	0.23	1	0.37	2	2	1	4
	1	1	1.00	0.00	0.38	0.62	0.83	0.17	0.00	0.00	0.00	0.00	0.00	0.00	2	0.63	3	3	1	2
58					4	5	1	2	1	2	0	1	0	1						
	2	2	1.00	0.00	0.54	0.46	0.92	0.08	0.00	0.00	0.71	0.29	0.26	0.74	2	0.06	2	2	2	4
	1	3	1.00	0.00	0.29	0.71	0.93	0.07	0.00	0.00	0.00	0.00	0.00	0.00	1	0.94	3	3	1	4
59					4	5	1	2	1	2	0	1	0	1						
	2	1	1.00	0.00	0.14	0.86	0.67	0.33	0.00	0.00	0.67	0.33	0.73	0.27	1	0.68	4	1	1	3
	1	2	1.00	0.00	0.82	0.18	0.81	0.19	0.00	0.00	0.00	0.00	0.00	0.00	2	0.32	3	1	1	2
60					3	5	1	2	1	2	0	1	0	1						
	2	2	0.37	0.63	0.11	0.89	0.82	0.18	0.86	0.14	0.24	0.76	0.72	0.28	1	0.31	3	2	2	2
	2	2	0.20	0.80	0.10	0.90	0.38	0.62	0.54	0.46	0.82	0.18	0.87	0.13	2	0.69	4	1	2	2
61					2	3	1	2	1	2	0	1	0	1						
	2	1	1.00	0.00	0.27	0.73	0.71	0.29	0.00	0.00	0.93	0.07	0.94	0.06	1	0.97	2	1	1	1
	1	1	1.00	0.00	0.33	0.67	0.95	0.05	0.00	0.00	0.00	0.00	0.00	0.00	2	0.03	3	1	1	1
62					2	3	1	2	1	2	0	1	0	1						
	2	1	1.00	0.00	0.47	0.53	0.82	0.18	0.00	0.00	0.99	0.01	0.95	0.05	2	0.12	3	1	1	2
	1	1	1.00	0.00	0.42	0.58	0.83	0.17	0.00	0.00	0.00	0.00	0.00	0.00	1	0.88	2	1	1	2
63					2	3	1	2	1	2	0	1	0	1						
	2	1	1.00	0.00	0.18	0.82	0.93	0.07	0.00	0.00	0.83	0.17	0.50	0.50	2	0.63	3	1	1	2
	1	1	1.00	0.00	0.05	0.95	0.97	0.03	0.00	0.00	0.00	0.00	0.00	0.00	1	0.37	2	2	1	2
64					2	3	1	2	1	2	0	1	0	1						
	2	1	1.00	0.00	0.06	0.94	0.78	0.22	0.00	0.00	0.94	0.06	0.82	0.18	2	0.32	2	2	1	1
	1	1	1.00	0.00	0.66	0.34	0.99	0.01	0.00	0.00	0.00	0.00	0.00	0.00	1	0.68	2	2	1	3

No.	N/M	K_i	p_{ij}		a_i		b_{i1}		b_{i2}		c_{j1}		c_{j2}		PN_i	$\tilde{p}_l$	TW_i	MN_i	LV_1	LV_2
65					2	3	1	2	1	2	0	1	0	1						
	2	1	1.00	0.00	0.22	0.78	0.84	0.16	0.00	0.00	0.77	0.23	0.73	0.27	2	0.36	1	1	1	3
	1	1	1.00	0.00	0.07	0.93	0.99	0.01	0.00	0.00	0.00	0.00	0.00	0.00	1	0.64	3	2	1	1
66					2	3	1	2	1	2	0	1	0	1						
	2	1	1.00	0.00	0.00	1.00	0.60	0.40	0.00	0.00	0.91	0.09	0.83	0.17	1	0.86	3	1	1	3
	1	1	1.00	0.00	0.04	0.96	0.82	0.18	0.00	0.00	0.00	0.00	0.00	0.00	2	0.14	3	1	1	3
67					2	3	1	2	1	2	0	1	0	1						
	2	1	1.00	0.00	0.01	0.99	0.46	0.54	0.00	0.00	0.95	0.05	0.88	0.12	1	0.78	1	1	1	3
	1	1	1.00	0.00	0.38	0.62	0.94	0.06	0.00	0.00	0.00	0.00	0.00	0.00	2	0.22	1	1	1	1
68					2	3	1	2	1	2	0	1	0	1						
	2	1	1.00	0.00	0.04	0.96	0.57	0.43	0.00	0.00	0.57	0.43	0.99	0.01	1	0.01	3	1	1	1
	1	1	1.00	0.00	0.01	0.99	0.93	0.07	0.00	0.00	0.00	0.00	0.00	0.00	2	0.99	3	1	1	2
69					2	3	1	2	1	2	0	1	0	1						
	2	1	1.00	0.00	0.55	0.45	0.90	0.10	0.00	0.00	0.89	0.11	0.88	0.12	2	0.25	2	2	1	3
	1	1	1.00	0.00	0.17	0.83	0.87	0.13	0.00	0.00	0.00	0.00	0.00	0.00	1	0.75	3	1	1	2
70					2	3	1	2	1	2	0	1	0	1						
	2	1	1.00	0.00	0.34	0.66	0.89	0.11	0.00	0.00	0.72	0.28	0.91	0.09	2	0.00	1	2	1	3
	1	1	1.00	0.00	0.17	0.83	0.74	0.26	0.00	0.00	0.00	0.00	0.00	0.00	1	1.00	1	2	1	1
71					2	3	1	2	1	2	0	1	0	1						
	2	1	1.00	0.00	0.01	0.99	0.92	0.08	0.00	0.00	0.65	0.35	0.88	0.12	1	0.01	1	2	1	2
	1	1	1.00	0.00	0.14	0.86	0.96	0.04	0.00	0.00	0.00	0.00	0.00	0.00	2	0.99	2	2	1	3
72					2	3	1	2	1	2	0	1	0	1						
	2	1	1.00	0.00	0.09	0.91	0.81	0.19	0.00	0.00	0.98	0.02	0.85	0.15	1	0.85	2	1	1	1
	1	1	1.00	0.00	0.01	0.99	0.80	0.20	0.00	0.00	0.00	0.00	0.00	0.00	2	0.15	2	2	1	2

No.	N/M	K_i	p_{ij}		a_i		b_{i1}		b_{i2}		c_{j1}		c_{j2}		PN_i	$\tilde{p}_l$	TW_i	MN_i	LV_1	LV_2
73					2	3	1	2	1	2	0	1	0	1						
	2	1	1.00	0.00	0.44	0.56	0.99	0.01	0.00	0.00	0.92	0.08	1.00	0.00	1	0.63	2	1	1	3
	1	1	1.00	0.00	0.32	0.68	0.90	0.10	0.00	0.00	0.00	0.00	0.00	0.00	2	0.37	3	2	1	2
74					2	3	1	2	1	2	0	1	0	1						
	2	1	1.00	0.00	0.11	0.89	0.91	0.09	0.00	0.00	0.86	0.14	0.87	0.13	1	0.93	3	1	1	1
	1	1	1.00	0.00	0.25	0.75	0.89	0.11	0.00	0.00	0.00	0.00	0.00	0.00	2	0.07	1	1	1	1
75					2	3	1	2	1	2	0	1	0	1						
	2	1	1.00	0.00	0.23	0.77	0.92	0.08	0.00	0.00	0.73	0.27	0.95	0.05	1	0.25	1	1	1	2
	1	1	1.00	0.00	0.26	0.74	0.95	0.05	0.00	0.00	0.00	0.00	0.00	0.00	2	0.75	2	1	1	3
76					2	3	1	2	1	2	0	1	0	1						
	2	1	1.00	0.00	0.20	0.80	0.97	0.03	0.00	0.00	0.99	0.01	0.80	0.20	2	0.39	1	1	1	2
	1	1	1.00	0.00	0.05	0.95	0.72	0.28	0.00	0.00	0.00	0.00	0.00	0.00	1	0.61	2	2	1	1
77					2	3	1	2	1	2	0	1	0	1						
	2	1	1.00	0.00	0.18	0.82	0.86	0.14	0.00	0.00	0.84	0.16	0.90	0.10	1	0.95	2	2	1	3
	1	1	1.00	0.00	0.37	0.63	0.95	0.05	0.00	0.00	0.00	0.00	0.00	0.00	2	0.05	2	2	1	2
78					2	3	1	2	1	2	0	1	0	1						
	2	1	1.00	0.00	0.13	0.87	0.70	0.30	0.00	0.00	0.87	0.13	0.97	0.03	2	0.19	2	2	1	2
	1	1	1.00	0.00	0.11	0.89	0.94	0.06	0.00	0.00	0.00	0.00	0.00	0.00	1	0.81	2	2	1	2
79					2	3	1	2	1	2	0	1	0	1						
	2	1	1.00	0.00	0.43	0.57	0.94	0.06	0.00	0.00	0.93	0.07	0.98	0.02	1	0.20	1	2	1	3
	1	1	1.00	0.00	0.37	0.63	0.85	0.15	0.00	0.00	0.00	0.00	0.00	0.00	2	0.80	2	2	1	2
80					2	3	1	2	1	2	0	1	0	1						
	2	1	1.00	0.00	0.13	0.87	0.75	0.25	0.00	0.00	0.95	0.05	0.82	0.18	1	0.48	2	1	1	1
	1	1	1.00	0.00	0.24	0.76	0.99	0.01	0.00	0.00	0.00	0.00	0.00	0.00	2	0.52	2	1	1	1

No.	N/M	K_i	p_{ij}		a_i		b_{i1}		b_{i2}		c_{j1}		c_{j2}		PN_i	$\tilde{p}_l$	TW_i	MN_i	LV_1	LV_2
81					2	3	1	2	1	2	0	1	0	1						
	2	1	1.00	0.00	0.05	0.95	0.98	0.02	0.00	0.00	0.99	0.01	0.99	0.01	2	0.70	2	1	1	3
	1	1	1.00	0.00	0.03	0.97	0.98	0.02	0.00	0.00	0.00	0.00	0.00	0.00	1	0.30	1	1	1	1
82					2	3	1	2	1	2	0	1	0	1						
	2	1	1.00	0.00	0.01	0.99	0.96	0.04	0.00	0.00	0.99	0.01	1.00	0.00	1	0.79	2	2	1	1
	1	1	1.00	0.00	0.04	0.96	0.98	0.02	0.00	0.00	0.00	0.00	0.00	0.00	2	0.21	1	2	1	3
83					2	3	1	2	1	2	0	1	0	1						
	2	1	1.00	0.00	0.05	0.95	0.99	0.01	0.00	0.00	0.99	0.01	0.99	0.01	2	0.75	2	1	1	3
	1	1	1.00	0.00	0.01	0.99	1.00	0.00	0.00	0.00	0.00	0.00	0.00	0.00	1	0.25	2	1	1	1
84					2	3	1	2	1	2	0	1	0	1						
	2	1	1.00	0.00	0.07	0.93	1.00	0.00	0.00	0.00	0.99	0.01	0.96	0.04	2	0.32	3	1	1	2
	1	1	1.00	0.00	0.01	0.99	1.00	0.00	0.00	0.00	0.00	0.00	0.00	0.00	1	0.68	3	1	1	2
85					2	3	1	2	1	2	0	1	0	1						
	2	1	1.00	0.00	0.09	0.91	1.00	0.00	0.00	0.00	0.97	0.03	1.00	0.00	1	0.46	3	1	1	3
	1	1	1.00	0.00	0.06	0.94	0.98	0.02	0.00	0.00	0.00	0.00	0.00	0.00	2	0.54	1	2	1	2
86					2	3	1	2	1	2	0	1	0	1						
	2	1	1.00	0.00	0.01	0.99	0.99	0.01	0.00	0.00	0.96	0.04	0.97	0.03	1	0.01	1	1	1	3
	1	1	1.00	0.00	0.01	0.99	0.99	0.01	0.00	0.00	0.00	0.00	0.00	0.00	2	0.99	2	1	1	2
87					2	3	1	2	1	2	0	1	0	1						
	2	1	1.00	0.00	0.01	0.99	0.98	0.02	0.00	0.00	0.98	0.02	0.96	0.04	2	0.80	2	1	1	3
	1	1	1.00	0.00	0.01	0.99	0.99	0.01	0.00	0.00	0.00	0.00	0.00	0.00	1	0.20	2	1	1	2
88					2	3	1	2	1	2	0	1	0	1						
	2	1	1.00	0.00	0.04	0.96	0.97	0.03	0.00	0.00	0.99	0.01	0.99	0.01	2	0.70	2	1	1	1
	1	1	1.00	0.00	0.03	0.97	0.99	0.01	0.00	0.00	0.00	0.00	0.00	0.00	1	0.30	2	1	1	3

No.	N/M	K_i	p_{ij}		a_i		b_{i1}		b_{i2}		c_{j1}		c_{j2}		PN_i	$\tilde{p}_l$	TW_i	MN_i	LV_1	LV_2
89					2	3	1	2	1	2	0	1	0	1						
	2	1	1.00	0.00	0.02	0.98	1.00	0.00	0.00	0.00	0.98	0.02	0.98	0.02	1	0.86	2	1	1	2
	1	1	1.00	0.00	0.10	0.90	0.99	0.01	0.00	0.00	0.00	0.00	0.00	0.00	2	0.14	3	1	1	2
90					2	3	1	2	1	2	0	1	0	1						
	2	1	1.00	0.00	0.03	0.97	1.00	0.00	0.00	0.00	0.99	0.01	0.95	0.05	2	0.69	3	1	1	1
	1	1	1.00	0.00	0.01	0.99	0.98	0.02	0.00	0.00	0.00	0.00	0.00	0.00	1	0.31	1	2	1	2

Table C.1: Table of randomly generated parameter configurations within the parameter limits of Table 6.1 for validating the analytical model by comparison with a simulation model

C.2 Table of the Basic Parameter Configurations for Numerical Evaluation

Category	Parameters	Notation	Value
System	Number of queues	N	5
	Number of sinks	M	5
	Queue capacities per queue	K_i	100
Transition	Distribution of the probability across the sinks	$D_{\hat{P}}$	uniform
	Gradient of the probability across the sinks	$b_{\hat{P}}$	0
	Highest probability for the transition from a queue to the sink with the same number	BV_1	0
Interarrival times	Type of distribution	T_A	gamma
	Expected value	$E(\tilde{A})$	11
	Variability of the distribution per queue	$c^2(A_i)$	1
	Distribution of the expected value across the queues	D_A	uniform
	Gradient of the expected value across the queues	b_A	0
Processing times	Type of distribution	T_S	gamma
	Expected value	$E(\tilde{S})$	5
	Variability of the distribution per queue and sink	$c^2(S_{ij})$	1
	Distribution of the expected value across the queues	D_{S_i}	uniform
	Gradient of the expected value across the queues	b_{S_i}	0
	Distribution of the expected value across the sinks	D_{S_j}	uniform
	Gradient of the expected value across the sinks	b_{S_j}	0
Switching times	Type of distribution	T_C	gamma
	Expected value	$E(\tilde{C})$	5
	Variability of the distribution per sink and queue	$c^2(C_{ji})$	1
	Distribution of the expected value across the queues	D_{C_i}	uniform
	Gradient of the expected value across the queues	b_{C_i}	0
	Distribution of the expected value across the sinks	D_{C_j}	uniform
	Gradient of the expected value across the sinks	b_{C_j}	0
	No switching time when switching from a sink to the queue with the same number	BV_2	0
Service rule	Type of priority distribution	D_{PN}	ascending
	Type of random distribution	$D_{\tilde{P}}$	uniform
	Time window time per queue	TW_i	14
	Maximum number of services per queue	MN_i	2
	Limit value of the queue length per queue	LV_i^1	66
	Limit value of the waiting time per queue	LV_i^2	3100
	Service rule number	SR	2

Table C.2: Basic setting of the input parameters for the numerical evaluation of the MQSMDS